Respiratory Biomechanics

Mary A. Farrell Epstein   James R. Ligas
Editors

Respiratory Biomechanics

Engineering Analysis of Structure and Function

With 79 Illustrations

Springer-Verlag New York Berlin Heidelberg
London Paris Tokyo Hong Kong

Mary A. Farrell Epstein
University of Connecticut Health Center
Department of Pharmacology
Farmington, CT 06032
USA

James R. Ligas
University of Connecticut
Department of Surgery
Farmington, CT 06032
USA

Printed on acid-free paper.

Softcover reprint of the hardcover 1st edition 1990

Camera-ready copy prepared by authors.

9 8 7 6 5 4 3 2 1

ISBN-13: 978-1-4612-8017-0 e-ISBN-13: 978-1-4612-3452-4

DOI: 10.1007/978-1-4612-3452-4

To Our Parents and Our Families

Preface

This proceedings volume brings together the invited papers from the Respiratory Biomechanics Symposium of the First World Congress of Biomechanics held in La Jolla, California from August 30–September 4, 1990. The respiratory system offers many opportunities to apply the different branches of traditional mechanics. Tissue deformations and stresses during lung expansion can be analyzed using the principles of solid mechanics. Fluid mechanical problems in the lung are unique. There is the matched distribution of two fluids, gas and blood, in two beautifully intertwined, branched conduit systems. The reversing flow of the gas phase presents different problems than the pulsatile flow of the non-Newtonian fluid that is the blood. On the smaller scale, there is the flux of fluids and solutes across the capillary membrane. Finally, there is the problem of coupling fluid and solid mechanics to understand the overall behavior of the respiratory system.

In this symposium, we have chosen to address the basic processes that contribute to the gas and fluid exchange functions of the lung. Section 1, Lung Tissue Mechanics, provides an historical background and, then, presents more recent work on the structure of the lung parenchyma, the mechanics of the tissue, and the effects of the bounding membrane, the visceral pleura.

In Section 2, Respiratory Fluid Mechanics and Transport, studies of gas flow and species transport are discussed. Respiratory impedance measurements provide non-invasive methods for assessing the ease of distribution of gas in the lung. The first two studies report the use of linear models to analyze impedance measurements during breathing, and advanced modeling approaches which yield guidelines for optimal design of experimental protocols. Next, geometric and mechanical factors affecting the transport of gaseous species in the conducting airways and at the alveolar-capillary level are presented. Last, the fluid mechanics of the nasal passage, a relatively new area in respiratory mechanics, is analyzed.

The third and fourth sections represent the most rapidly expanding areas in respiratory system mechanics: fluid mechanics and transport in the pulmonary vasculature. In Section 3, Pulmonary Circulation, the mechanical properties of pulmonary blood vessels and capillaries are described. Then, methods for determining capillary recruitment, pulsatile pulmonary pressure, and sites of vasoconstriction are presented. Section 4, The Use of Mathematics and Advanced Technology to Measure and Evaluate Lung Fluid Exchange and Solute Balance, continues with methods for measuring vascular permeability, protein transport, and endothelial barrier function. Mathematical approaches discussed include models of solute exchange across the epithelium and fractal analysis of lung fluid flux.

The fifth and last section, Integrating Mechanics and Transport in Assessing Respiratory Function, seeks to integrate the basic physical and molecular mechanisms presented in the first four sections by focusing on three major processes in the lung: airway heat and water exchange, pulmonary blood flow distribution, and gas exchange.

The editors wish to thank the organizing committee of the World Congress for the opportunity to bring distinguished investigators together at an international forum to present their analyses of current work and delineate future areas for research. We especially wish to thank Dr. Savio Woo, Program Chairman, and Drs. Shu Chien, Y.C. Fung, Geert Schmid-Schönbein, and Richard Skalak for their continued encouragement during development of this Symposium. We also appreciate the efforts of our Symposium session chairs, Drs. Daniel Isabey, John Linehan, Thomas Harris, Luis Oppenheimer, and Michael Hlastala, who worked actively with us to invite an outstanding group of investigators to participate in this Symposium. Special thanks go to all our authors who found time at the end of the academic year and in the midst of grant preparation and attendance at professional meetings to prepare their manuscripts.

The editors are grateful for the expert and timely assistance of Dr. Zvi Ruder, Engineering Editor of Springer-Verlag, and his staff which greatly facilitated work on this volume. Our special thanks go to Ms. Andrea Tedesco who, with constant good humor and conscientious care, provided organizational support of correspondence and manuscripts throughout this project.

We hope that this volume will be a source of new viewpoints and encouragement for those engaged in or considering research in respiratory system biomechanics.

May, 1990

Mary A. Farrell Epstein, Eng.Sc.D.
James R. Ligas, M.D., Ph.D.

University of Connecticut
Medical School
Farmington, CT 06032

Contents

3. Pulmonary Circulation

4. The Use of Mathematics and Advanced Technology to Measure and Evaluate Lung Fluid Exchange and Solute Balance

5. Integrating Mechanics and Transport in Assessing Respiratory Function

Contributors

Valerie J. Abernathy
Department of Biomedical Engineering,
Vanderbilt University,
Nashville, TN 37235, USA

H.K. Chang
Department of Biomedical Engineering,
University of Southern California,
Los Angeles, CA 90089-1451, USA

Michael L. Collins
Department of Physiology,
University of Texas Health Center,
Tyler, TX 75710, USA

Jon C. Connelly
Department of Physiology,
University of Texas Health Center,
Tyler, TX 75710, USA

Christopher A. Dawson
Department of Biomedical Engineering,
Marquette University,
Milwaukee, WI 53233, USA

Mary A. Farrell Epstein
Department of Pharmacology,
University of Connecticut Health Center,
Farmington, CT 06032, USA

William J. Federspiel
Department of Biomedical Engineering,
Boston University,
Boston, MA 02215 USA

Elena Furuya
Respiratory Investigation Unit,
University of Manitoba,
Winnipeg, Manitoba,
Canada R3A 1R9

Robb W. Glenny
Department of Pulmonary and Critical Care Medicine,
University of Washington,
Seattle, WA 98195, USA

Lynn D. Gray
Department of Cell Biology,
University of Texas Health Center,
Tyler, TX 75710, USA

James B. Grotberg
Department of Biomedical Engineering,
Northwestern University,
Evanston, IL 60208, USA

Thomas R. Harris
Department of Biomedical Engineering,
Vanderbilt University,
Nashville, TN 37235, USA

Michael P. Hlastala
Department of Pulmonary and Critical Care Medicine,
University of Washington,
Seattle, WA 98195, USA

David Holiday
Department of Biomathematics,
University of Texas Health Center,
Tyler, TX 75710, USA

Don Huebert
Respiratory Investigation Unit,
University of Manitoba,
Winnipeg, Manitoba,
Canada R3A 1R9

Daniel Isabey
INSERM Unité 298,
Faculté de Médecine de Créteil,
94010 Créteil, France

Andrew C. Jackson
Department of Biomedical Engineering, Boston University,
Boston, MA 02215, USA

James D. Kaplan
Respiratory and Critical Care Division, Washington University School of Medicine,
St. Louis, MO 63110, USA

Stephen J. Lai-Fook
Department of Biomedical Engineering,
University of Kentucky,
Lexington, KY 40506-0070, USA

Kevin P. Landolfo
Respiratory Investigation Unit,
University of Manitoba,
Winnipeg, Manitoba,
Canada R3A 1R9

J.S. Lee
Department of Biomedical Engineering,
University of Virginia,
Charlottesville, VA 22908, USA

L.P. Lee
Department of Biomedical Engineering,
University of Virginia,
Charlottesville, VA 22908, USA

James R. Ligas
Departments of Surgery and Anesthesiology,
University of Connecticut Health Center,
Farmington, CT 06032, USA

John H. Linehan
Department of Biomedical Engineering,
Marquette University, Milwaukee, WI 53233, USA

B. Louis
INSERM Unité 298,
Faculté de Médecine de Créteil,
94010, Créteil, France

Jean-Michel Maarek
Department of Biomedical Engineering,
University of Southern California,
Los Angeles, CA 90089-1451, USA

Joanne Markham
Institute of Biomedical Computing,
Washington University School of Medicine,
St. Louis, MO 63110, USA

Jerry W. McLarty
Department of Biomathematics,
University of Texas Health Center,
Tyler, TX 75710, USA

James E. McNamee
Department of Physiology,
University of South Carolina
School of Medicine,
Columbia, SC 29208, USA

Robert R. Mercer
Department of Medicine, Duke
University,
Durham, NC 22710, USA

Mark Mintun
Division of Nuclear Medicine,
University of Michigan Medical
Center,
Ann Arbor, MI 48109-0048, USA

E.H. Oldmixon
Department of Medicine,
Memorial Hospital of Rhode
Island, Pawtucket, RI 02860, USA

Luis Oppenheimer
Respiratory Investigation Unit,
University of Manitoba,
Winnipeg, Manitoba,
Canada R3A 1R9

Richard E. Parker
Department of Radiology,
Vanderbilt University,
Nashville, TN 37235, USA

Barry T. Peterson
Department of Physiology,
University of Texas Health Center,
Tyler, TX 75710, USA

N. Adrienne Pou
Department of Biomedical
Engineering, Vanderbilt
University, Nashville, TN 37235,
USA

William R. Riddle
Center for Lung Research,
Vanderbilt University,
Nashville, TN 37235, USA

H. Thomas Robertson
Department of Pulmonary and
Critical Care Medicine,
University of Washington,
Seattle, WA 98195, USA

Z. Rong
Shanghai Medical University,
P.R. of China

Robert J. Roselli
Department of Biomedical
Engineering,
Vanderbilt University,
Nashville, TN 37235, USA

Daniel P. Schuster
Respiratory and Critical Care
Division,
Washington University School of
Medicine,
St. Louis, MO 63110, USA

Julian Solway
Department of Medicine,
The University of Chicago,
Chicago, IL 60637, USA

Kevin J. Sullivan
Department of Biomedical
Engineering,
University of Southern California,
Los Angeles, CA 90089-1451,
USA

D. Tai
Department of Radiology,
University of Tennessee,
Memphis, TN 38163, USA

James S. Ultman
Department of Chemical
Engineering,
Pennsylvania State University,
University Park, PA 16802,
USA

Peter D. Wagner
Department of Medicine,
University of California,
San Diego, La Jolla, CA 92093,
USA

Wiltz W. Wagner, Jr.
Departments of Anesthesiology
and Physiology-Biophysics,
Indiana University Medical School,
Indianapolis, IN 46223, USA

Thomas Warfel
Medical Student Office,
University of Pittsburgh,
Pittsburgh, PA 15261, USA

R.T. Yen
Department of Biomedical
Engineering,
Memphis State University,
Memphis, TN 38152, USA

B. Zhang
Shanghai Medical University,
P.R. of China

LUNG TISSUE MECHANICS

Lung Tissue Mechanics: Historical Overview

James R. Ligas, University of Connecticut Health Center, Farmington CT 06032

In this brief an article it is impossible to critically review the many papers on the solid mechanics of the lung parenchyma. I shall therefore present a framework into which investigations may be placed and provide a somewhat restricted bibliography.

Pressure-volume behavior of the lung:

Early experiments on lung mechanical behavior focused on simple measures which might relate to tissue stresses and strains: the distending pressure difference applied to an excised lung and the resulting change in volume. These curves showed hysteresis, stress relaxation, and creep. They differed markedly if the lung were degassed and then inflated with and surrounded by saline (Figure 1). The hysteresis area was reported to be independent of lung volume or frequency of cycling [3, 4, 27, 29]. Although most of the time-dependent behavior and hysteresis were eliminated by saline filling, some remained.

Radford [61] suggested that three mechanisms could contribute to hysteresis: the surfactant lining the alveoli, geometric irreversibility in the number of open lung units, and nonelastic behavior of the tissues themselves. Each received some degree of investigation which ultimately had important implications for early continuum analyses of lung mechanics.

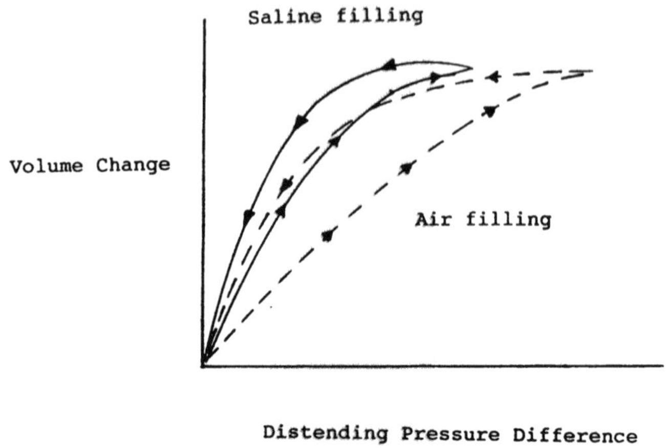

Volume-Pressure Curves with Air and Saline Filling
Arrows denote inflation and deflation

FIGURE 1

Clements [10] investigated the surface active agent lining the alveoli at the air/liquid interface. He determined the surface tension of lung extracts floated on saline in a Wilhelmy balance. Surface tension varied with the surface area of the film. More recent studies of surfactant have been made by comparing pressure-volume data from air- to liquid-filled lungs [67, 83], by using a new captive bubble surface tensiometer [66], or by studying shape changes in subpleural alveoli to ascertain the effects of surface forces [15]. There have also been attempts to directly gauge the effects of surfactant at the alveolar level [64,65]. All of these studies reach the same conclusion: that surface tension at the air/liquid interface in the alveoli plays an important role in lung mechanics. However, mathematical models for surfactant [1, 33, 34, 83] are complex and difficult to validate. Some indicate that in addition to providing a surface tension which varies with area, surfactant may be responsible for some of the time-dependent effects as well [33].

Thus one must be cautious in interpreting experiments performed in air-filled lungs as arguments for time-dependent tissue behavior [27, 28, 29]. To avoid the complexities associated with surfactant, many continuum models for the respiratory system employed data taken from saline-inflated lungs to try to analyze tissue properties alone. Others relied upon data from air-filled lungs, assuming that the effects of surfactant can be incorporated into the mechanical properties of the tissue. Thus the reader must be careful to examine both the model and the data from which the reputed mechanical properties were calculated.

Geometric irreversibility was also investigated [2, 12, 14, 16, 17, 24, 30, 31, 39, 60, 75, 86]. Many of these studies were performed in intact animals. The lungs were expanded to a given volume, the animal sacrificed, and the lungs fixed and examined. Morphologic data suggested two possible mechanisms for hysteresis based on geometric changes: (1) the number of expanded alveoli can vary: at low lung volumes some alveoli collapse and higher pressures are required for subsequent reinflation; and (2) the alveolar walls may fold or pleat at low lung volumes. Although the data are conflicting, most investigators concluded that at lung volumes above functional residual capacity, geometric irreversibility plays a minor role, if any, in lung mechanics. Accordingly, many continuum models for the lung postulate a fixed number of alveoli of a certain shape which are deformed by subsequent lung expansion. At no point in the inflation/deflation cycle are drastic shape changes allowed to occur.

The residual hysteresis and time-dependent effects noted in saline-filled lungs were recognized early on as a possible indicator of viscoelastic or plastic tissue behavior [27, 28, 29, 35, 52]. However, because the added stresses due to these effects appeared to be small, most continuum models treat the lung parenchyma as if it were purely elastic.

Thus early observations taken from intact or excised whole lung pressure-volume data together with some structural studies and surface tension experiments led to an approximation of lung mechanical behavior by models which did not incorporate geometric irreversibility, were based on elasticity theory, and for the most part did not attempt to deal with the complexities of surface active agents at air/liquid interfaces.

Continuum analyses of the lung:

Continuum analyses require an equation of motion, a relationship between stress and strain for the material, and some boundary conditions relating the applied loads to the resulting stresses and strains.

Most investigations of lung mechanics have relied on steady state data to model pressure-volume relationships so that the equation of motion reduces to a static force balance.

The first requirement for developing a constitutive, or stress-strain, relationship is that one must be able to define a reference state from which deformations can be measured. Fortunately, the lung does seem to return to a reproducible volume when all loads are removed [77]. Thus deformations are definable and one can attempt to relate those deformations to the applied stresses. In spite of the fact that the data cited above suggested that the lung may be a viscoelastic or plastic tissue, investigations of time-dependent phenomena were few in number compared to experiments designed to measure lung elasticity.

Stress-strain data:

Although there are some data on regional deformations in intact animals [7,8] for the most part our knowledge of the mechanical properties of lung tissues arises from in vitro experiments. In addition to the pressure-volume experiments, data were taken from small samples of alveolar wall [20, 72, 73] or from tissue blocks loaded in tension and compression [32, 78, 74]. Most experiments focused on steady-state behavior. For the purpose of developing a constitutive, or stress-strain, relationship two questions were of interest: is the material of which the lung is composed isotropic, and what is the form of a suitable equation to describe the relationship?

Most studies imply that the assumption of homogeneity and isotropy for the lung parenchyma is reasonable, at least as a first approach for normal lungs [74]. Furthermore, there seems to be a roughly exponential relationship between stress and strain:

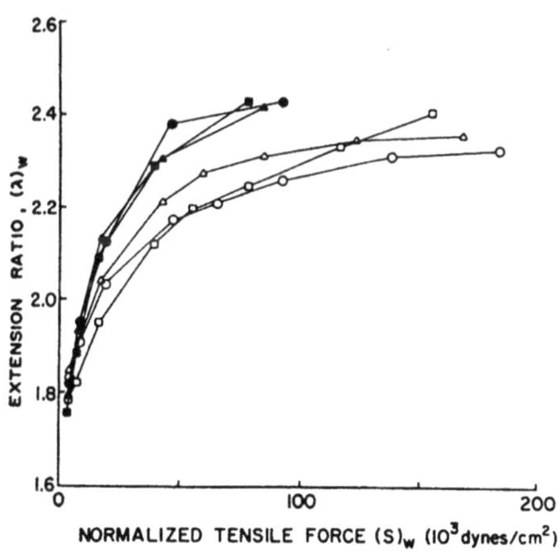

Applied stresses versus strains:
cubic blocks of lung tissue loaded triaxially.
From [32].

FIGURE 2

Although time-dependent behaviors were generally ignored, some attempt to incorporate hysteresis was made by treating the lung tissue as pseudoelastic. This term was used to imply that although the relationship of stress to strain, and hence the presumed material properties, differs between inhalation and exhalation, in each one of these phases behavior is perfectly elastic.

Constitutive relationships:

One can divide the majority of papers into four categories based upon two criteria:

(1) A macroscopic versus a microscopic focus
(2) Finite versus infinitesimal displacements

By a macroscopic focus I refer to those works which treat the lung as a spongy continuum with bulk mechanical properties which may well reflect the geometric arrangement of the alveolar walls. By a microscopic focus I refer to attempts to describe the mechanical properties of the alveolar membranes and couple that to an alveolar geometry to arrive at the behavior of bulk lung tissue.

Once again, one must be careful when reviewing these papers to determine whether the effects of surfactant are incorporated into the mechanical properties.

Macroscopic approach:

Several constitutive relationships have been proposed, all of which are pseudoelastic. Some apply to finite deformations, while others apply to infinitesimally small deformations only. The latter are linearized theories generally employing the elastic or bulk moduli, which are allowed to depend upon lung volume or the level of the distending pressure [41, 42, 43, 44, 47, 55, 71, 82]. As Lee [47] points out, this use of incremental isotropic laws is valid only when one can superimpose a small nonuniform deformation on a large isotropic strain pattern.

Microscopic approach:

There are several proposed constitutive equations for the alveolar membranes themselves, and again most are pseudoelastic. The geometries applied to the analyses range from cubic alveoli to spherical alveoli to dodecahedrons, and the constitutive equations range from linear small-displacement theory to exponential forms in terms of the strain invariants [18, 21, 46, 49, 79]. To arrive at the behavior of bulk lung tissue from the mechanical properties and geometry of the membranes themselves, Fung [21] formulated what he called the Ergodic Hypothesis: that macroscopically, at a point within the lung tissue the net effect of averaging the stresses over a large number of alveoli cut by a single plane was equal to the average over a single alveolus cut by a large number of planes. This technique of specifying how microstructures such as alveolar walls are distributed near a point in the macroscopic lung remains basic to the microscopic approach. Lanir [45] lent it a new formalism when he essentially made the density distribution of the spatial orientation of alveolar membranes a structural parameter of his model.

Structural studies:

Both for the microscopic approach and to understand mechanical phenomena at smaller scales than the behavior of bulk lung tissue, the exact microstructure of the alveolar duct is important. This was recognized early on

[62]. Many investigators have performed microstructural studies to determine the exact orientation of the alveolar membranes, the components of the membranes themselves, and the reinforcement of the membranes by collagen and elastin fibers to determine exactly how loads are carried at the microstructural level [9, 25, 26, 53, 56, 57, 58, 59, 68]. In addition, investigations of dynamic rearrangements using backscattered light from the pleural surface make it possible to obtain dynamic data, at least in terms of changes in gross geometric measurements [6]. Rodarte [63] expanded the techniques available to determine in vivo deformations [7,8] by performing CT scans on isolated dog lobes at different volumes to study variability of expansion in cubes 1.5 cm on a side.

These structural investigations have resulted in some rethinking of how lung structure bears loads [5, 38, 71], how idealized geometric figures may be applied to structural models [23], and how constitutive relationships may be determined by tissue components [11, 37,51, 69].

Applications:

Continuum analyses have been applied together with finite element methods to characterize the deformation of an entire lung or lobe loaded by gravity, the deformation about circular or cylindrical holes such as airways or blood vessels coursing through the parenchyma, or to reproduce the pressure-volume curve of the intact lung [13, 22, 36, 40, 42, 48, 54, 70, 76, 80, 81, 84]. Some of these have been reviewed elsewhere [85].

Most of the papers cited focus on our qualitative understanding of lung expansion. The ability of such models to discriminate between normal and diseased lungs, even qualitatively, has not been tested. Only one study [50] addresses the question of whether the parameter values obtained from such models are reproducible or consistent throughout a series of normal lungs. It suggests that a specific continuum model applied to pressure-volume data from normal excised canine lungs yields parameter values lying in a narrow range. Furthermore, most of the parameters seem to be fairly well-determined. However, there is no such study performed upon abnormal lungs for comparison.

Conclusions:

The papers above for the most part rely upon in vitro data on tissue mechanical behavior. In addition to the question of whether living tissues are mechanically similar to freshly excised tissue, in vivo lungs are affected by vascular perfusion, surfactant, and the effects of the chest wall. As our ability to make in vivo measurements of deformation fields improves, we may discover a need for more sophisticated models incorporating these effects. In addition, there is a resurgence of interest in the departure of lung tissue from perfect elasticity [67] and in understanding the relationship between mechanical behavior and structure at the alveolar level.

References:

1. Archie JP Jr: An analytic evaluation of a mathematical model for the effect of pulmonary surfactant on respiratory mechanics. Dis Chest 53:759, 1968.
2. Ardila R, Horie T, Hildebrandt J: Macroscopic isotropy of lung expansion. Resp Physiol 20:105-115, 1974.
3. Bachofen H, Hildebrandt J: Area analysis of pressure-volume hysteresis in mammalian lungs. J Appl Physiol 30:493, 1971.
4. Bayliss LE, Robertson GW: The viscoelastic properties of the lungs. Quant J Exptl Physiol 29:27-47, 1939.
5. Budiansky B, Kimmel E: Elastic moduli of lungs. J Appl Mech 54:351-358, 1987.
6. Butler JP, Miki H, Suzuki S, Takishima T: Step response of lung surface-to-volume ratio by light scattering stereology. J Appl Physiol 67:1873, 1989.
7. Chevalier PA, Greenleaf JA, Robb RA, Wood EH: Biplane videographic analysis of dynamic regional lung strains in dogs. J Appl Physiol 40:118, 1976.
8. Chevalier PA, Rodarte JR, Harris LD: Regional lung expansion at total lung capacity in intact vs excised canine lungs. J Appl Physiol 45:363, 1978.
9. Ciurea D, Gil J: Morphometric study of human alveolar ducts based on serial sections. J Appl Physiol 67:2512, 1989.
10. Clements JA: Surface phenomena in relation to pulmonary function. Physiologist 5:11, 1962.
11. Comminou M, Yannas IV: Dependence of stress-strain nonlinearity of connective tissues on the geometry of collagen fibers. J Biomech 9:427, 1976.
12. Daly BDT, Parks GE, Edmonds CH, Hibbs CW, Norman JC: Dynamic alveolar mechanics as studied by videomicroscopy. Resp Physiol 24:217, 1975.

13. DeWilde R, Clement J, Hellemans JM, Decramer M et al: Model of elasticity of the human lung. J Appl Physiol 51:254, 1981.
14. Dunnill MS: Effect of lung inflation on alveolar surface area in the dog. Nature 214:1013, 1967.
15. Flicker E, Lee JS: Equilibrium of force of subpleural alveoli: Implications to lung mechanics J Appl Physiol 36:366, 1974.
16. Forrest JB: The effect of changes in lung volume on the size and shape of alveoli. J Appl Physiol 210:533, 1970.
17. Forrest JB: Lung tissue plasticity: morphometric analysis of anisotropic strain in liquid filled lungs. Resp physiol 27:223, 1976.
18. Frankus A, Lee GC: A theory for distortion studies of lung parenchyma based on alveolar membrane properties. J Biomech 7:101, 1974.
19. Fredberg JJ, Stamenovic D: On the imperfect elasticity of lung tissue. J Appl Physiol 67:2408, 1989.
20. Fukaya H, Martin CJ, Young AC, Katsura S: Mechanical properties of alveolar walls. J Appl Physiol 25:689, 1968.
21. Fung YC: A theory of elasticity of the lung. J Appl Mech 41:8, 1974.
22. Fung YC: Stress, deformation, and atelectasis of the lung. Cir Res 37:481, 1975.
23. Fung YC: A model of the lung structure and its validation. J Appl Physiol 64:2132, 1988.
24. Gil J, Weibel ER: Morphological study of pressure-volume hysteresis in rat lungs fixed by vascular perfusion. Resp Physiol 15:190, 1972.
25. Hansen JE, Ampaya EP, Bryant GH, Navin JJ: Branching pattern of airways and air spaces of a single human terminal bronchiole. J Appl Physiol 38:983, 1975.

26. Hansen JE, Ampaya EP: Human air space shapes, sizes, areas, and volumes. J Appl Physiol 38:990, 1975.
27. Hildebrandt J: Dynamic properties of air-filled excised cat lung determined by liquid plethysmograph. J Appl Physiol 27:246, 1969.
28. Hildebrandt J: Comparison of mathematical models for cat lung and viscoelastic balloon derived by Laplace transform methods from pressure-volume data. Bull Math Biophys 31:651, 1969.
29. Hildebrandt J: Pressure-volume data of cat lung interpreted by a plastoelastic, linear viscoelastic model. J Appl Physiol 28:365, 1970.
30. Hills BA: Geometric irreversibility and compliance hysteresis in the lung. Resp Physiol 13:50, 1971.
31. Hills BA: Effects of DPL at mercury/water interfaces and estimation of lung surface area. J Appl Physiol 36:41, 1974.
32. Hoppin FG Jr, Lee GC, Dawson SV: Properties of lung parenchyma in distortion. J Appl Physiol 39:742, 1975.
33. Horn LW, Davis SH: Apparent surface tension hysteresis of a dynamical system. J Colloid Interface Sci 151:459, 1975.
34. Horn LW: Evaluation of some alternative mechanisms for interface-related stress relaxation in lung. Resp Physiol 34:345, 1978.
35. Hughes RA, May J, Widdicombe JG: Stress relaxation in rabbits' lungs. J Physiol 146:85, 1959.
36. Karakaplan AD, Bieniek MP, Skalak R: A mathematical model of lung parenchyma. J Biomech Engr 102:124, 1980.
37. Karlinsky JB, Bowers JT III, Fredette JV, Evans J: Thermoelastic properties of uniaxially deformed lung strips. J Appl Physiol 58:459, 1985.

38. Kimmel E, Kamm RD, Shapiro AH: Cellular model of lung elasticity. J Biomech Engr 109:126, 1987.
39. Klingele TG, Staub NC: Alveolar shape changes with volume in isolated, air-filled lobes of cat lung. J Appl Physiol 28:411, 1970.
40. Kowe R, Schroter RC, Matthews FL, Hitchings D: Analysis of elastic and surface tension effects in the lung alveolus using finite element methods. J Biomech 19:541, 1986.
41. Lai-Fook SJ, Wilson TA, Hyatt RE, Rodarte JR: Elastic constants of inflated lobes of dog lungs. J Appl Physiol 40:508, 1976.
42. Lai-Fook SJ: Lung parenchyma described as a prestressed compressible material. J Biomech 10:357, 1977.
43. Lai-Fook SJ, Hyatt RE, Rodarte JR: Elastic constants of trapped lung parenchyma. J Appl Physiol 44:853, 1978.
44. Lambert RK, Wilson TA: A model for the elastic properties of the lung and their effect on expiratory flow. J Appl Physiol 34:34, 1973.
45. Lanir Y: Constitutive equations for the lung tissue. J Biomech Engr 105:374, 1983.
46. Lee GC, Frankus A: Elasticity properties of lung parenchyma derived from experimental distortion data. Biophys. J 15:481, 1975.
47. Lee GC, Frankus A, Chen PD: Small distortion properties of lung parenchyma as a compressible continuum. J Biomech 9:641, 1976.
48. Lee GC, Tseng NT, Yuan YM: Finite element modeling of lungs including interlobar fissures and the heart cavity. J Biomech 16:679, 1983.
49. Ligas JR: A nonlinearly elastic, finite deformation analysis applicable to the static mechanics of excised lungs. J Biomech 17:549, 1984.

50. Ligas JR, Saidel GM, Primiano FP Jr: Parameter estimation and sensitivity analysis of a nonlinearly elastic static lung model. J Biomech Engr 107:315, 1985.
51. Maes M, Vanhuyse VJ, Decraemer WF, Raman ER: A thermodynamically consistent constitutive equation for the elastic force-length relation of soft biological materials. J Biomech 22:1203, 1989.
52. Marshall R, Widdicombe JG: Stress relaxation of the human lung. Clin Sci 20:19, 1960.
53. Matsuda M, Fung YC, Sobin SS: Collagen and elastin fibers in human pulmonary alveolar mouths and ducts. J Appl Physiol 63:1185, 1987.
54. Matthews FL, West JB: Finite element displacement analysis of a lung. J Biomech 5:591, 1972.
55. Mead J, Takishima T, Leith D: Stress distribution in lungs: a model of pulmonary elasticity. J Appl Physiol 28:596, 1970.
56. Mercer RR, Crapo JD: 3-Dimensional reconstruction of the rat acinus. J Appl Physiol 63:785, 1987.
57. Mercer RR, Laco JM, Crapo JD: 3-dimensional reconstruction of alveoli in the rat lung for pressure-volume relationships. J Appl Physiol 62:1480, 1987.
58. Oldmixon EH, Butler JP, Hoppin FG Jr: Lengths and topology of alveolar septal borders. J Appl Physiol 67:1930, 1989.
59. Oldmixon EH, Hoppin FG Jr: Distribution of elastin and collagen in canine lung alveolar parenchyma. J Appl Physiol 67:1941, 1989.
60. Pierce JA, Hocott JB, Hefley WF: Elastic properties and the geometry of the lungs. J Clin Invest 40:1515, 1961.
61. Radford EP: Influence of physicochemical properties of the pulmonary surfce on stability of alveolar air spaces and on static hysteresis of lungs. IN: XXII International Congress of Physiol. Sci.,

Symposia and Special Lectures, Leiden, XXII Intl. Congress Sci., I:275-289, 1962.
62. Reifenrath R: The significance of alveolar geometry and surface tension in the respiratory mechanics of the lung. Resp Physiol 24:115, 1975.
63. Rodarte JR, Chaniotakis M, Wilson TA: Variability of parenchymal expansion measured by computed tomography. J Appl Physiol 67:226, 1989.
64. Schurch S, Goerke J, Clements JA: Direct determination of surface tension in the lung. Proc Nat Acad Sci 73:4698, 1976.
65. Schurch S: Surface tension at low lung volumes: dependence on time and alveolar sizes. Resp Physiol 48:339, 1982.
66. Schurch S, Bachofen H, Goerke J, Possmayer F: A captive bubble method reproduces the in situ behavior of lung surfactant monolayers. J Appl Physiol 67:2389, 1989.
67. Smith JC, Stamenovic D: Surface forces in lungs. I. Alveolar surface tension- lung volume relationships. J Appl Physiol 60: 1341, 1986.
68. Sobin SS, Fung YC, Tremer HM: Collagen and elastin fibers in human pulmonary alveolar walls. J Appl Physiol 64:1659, 1988.
69. Soong TT, Huang WN: A stochastic model for biological tissue elasticity in simple elongation. J Biomech 6:451, 1973.
70. Stamenovic D, Wilson TA: A strain energy function for lung parenchyma. J Biomech Engr 107:81, 1985.
71. Stamenovic D, Yager D: Elastic properties of air- and liquid-filled lung parenchyma. J Appl Physiol 65:2565, 1988.
72. Sugihara T, Martin CJ, Hidebrandt J: Length-tension properties of alveolar wall in man. J Appl Physiol 30:874, 1971.
73. Sugihara T, Hildebrandt J, Martin CJ: Viscoelastic properties of alveolar wall. J Appl Physiol 33:93, 1972.

74. Tai RC, Lee GC: Isotropy and homogeneity of lung tissue deformation. J Biomech 14:243, 1981.
75. Tsunoda S, Fukaya H, Sugihara T, Martin CJ, Hildebrandt J: Lung volume, thickness of alveolar walls, and microscopic anisotropy of expansion. Resp Physiol 22:285, 1974.
76. Vawter DL, Matthews LF, West JB: Effect of shape and size of lung and chest wall on stresses in the lung. J Appl Phys 39:9, 1975.
77. Vawter DL: Stress-free equilibrium volume of the lung. J Appl Physiol 43:3, 1977.
78. Vawter DL, Fung YC, West JB: Elasticity of excised dog lung parenchyma. J Appl Physiol 45:261, 1978.
79. Vawter DL, Fung YC, West JB: Constitutive equation of lung tissue elasticity. J Biomech Engr 101:38, 1979.
80. Vawter DL: A finite element model for macroscopic deformation of the lung. J Biomech Engr 102:1, 1980.
81. West JB, Matthews FL: Stresses, strains, and surface pressures in the lung caused by its weight. J Appl Physiol 32:332, 1972.
82. Wilson TA: A continuum analysis of a two-dimensional mechanical model of the lung parenchyma. J Appl Physiol 33:472, 1972.
83. Wilson TA: Relations among recoil pressure, surface area, and surface tension in the lung. J Appl Physiol 50:921, 1981.
84. Wilson TA, Bachofen H: A model of mechanical structure of the alveolar duct. J Appl Physiol 52: 1064, 1982.
85. Wilson TA: Nonuniform lung deformations. J Appl Physiol 54:1443, 1983.
86. Young SL, Tierney DF, Clements JA: Mechanism of compliance change in excised rat lungs at low transpulmonary pressure. J Appl Physiol 29:780, 1970.

ARCHITECTURE OF LUNG PARENCHYMA E.H. Oldmixon, Departments of Medicine, Memorial Hospital of RI and Brown University, Pawtucket, Rhode Island

1. *Introduction*

Two essential tasks of the mammalian lung are to provide a large gas-exchanging surface and to promote both the supply of oxygen-rich air to this surface and the removal of waste gasses from it. The mammalian lung solves the problem of providing ample surface by being, in large part, an open-celled foam, and it accomplishes the supply and removal of gasses by passively changing its volume while under varying tension. Neither solution is the only one available to an organism; the exchanging surfaces of avian lungs are more similar to bundled tubes, and air is propelled through this constant-volume array by the action of peripheral sacs.

Our understanding of the mammalian lung has been a long time evolving. That a lung could be inflated and would expel air by contracting of its own accord must have been known to butchers for millennia, but our first record of a careful observation of the property of elastic retraction (recoil) of lungs dates from 1820, when Carson (1) measured an increase in air pressure in the trachea of a bullock when its chest was opened. More than establishing the potential of the lung to recoil, this experiment showed that the lungs were distended beyond their relaxed state. Our research group (F.G. Hoppin, J.P. Butler, present author, and others) has continued to ask related questions at finer structural levels ("What is the configuration of lung parenchyma in the living mammal?", "Where is mechanically significant material distributed within the lung?", "How are the mechanically significant elements interconnected?", and of course, "Why does it work?"). This short article shall consider two topics: first, the known and unknown properties and interconnections between the mechanically significant elements of parenchyma, and second, the effort to reconcile views of parenchyma provided by different imaging methodologies.

2. *The mechanically significant elements of parenchyma and their interconnections.*

If alveolar parenchyma is responsible for most lung recoil, (>75%, by some estimates) and alveolar septa contain the greatest fraction of mass in parenchymal tissue, then alveolar septa should be mechanically significant elements ("membranes"). For the moment, we shall not consider whether septa simply provide a platform for surface tension at the air-lining layer interface or whether they also exert tension due to tensed tissue. Since septa *in vivo* are stretched beyond their relaxed dimensions, and because in an open-celled foam some septal edges are not attached to other septa, some special element producing a countering stress must be present at least at septal free edges. As we know, concentrations of connective tissue rich in elastin ("cables") are to be found within free septal edges and at other locations associated with alveolar entrance rings (6). (We shall see that these cables are also associated with the entire width and depth of alveoli.) Parenchyma, then, may be characterized as a cable/membrane structure, where the cables are elastic, curved, tensed, and elastin-rich, and the membranes are essentially planar, at least exerting surface tension and probably exerting tissue tension at most lung volumes, and elastin-poor.

Three membranes may attach to each other along a line of junction (Fig. 1A). Four such junction lines, and thus six membranes, may meet within a small volume. In general, these small volumes where septal borders confluesce are abstracted to point-like features called "nodes".

Putting aside nodes and concentrating on linear features, we find five border types in all (Fig. 1A-E): the junction (three membranes), the free edge or end (a term reflecting this border's appearance when sectioned) (1 cable, 1 membrane), the bend (1 cable, 2 membranes), the respiratory bronchiolar attachment, and the fixed structure attachment (both involving some sturdier structure, e.g. respiratory bronchiolar wall, conducting airway, vessel the size of arteriole or venule and larger, and 1 membrane) (5).

2.1. *Alveolar septal borders.*

The commonest septal border, the three-way junction, where three septa meet along a common line, is probably straight. If the septa are planar near the junction, then it is certainly so. A border's shape where septa abut vessels or conducting airways is likely to be controlled by the profile of the sturdier structure to which the septum attaches.

There are other kinds of septal borders, in contrast, which must be curved. The free edges of septa are reinforced by elastic, elastin-rich cables which are tensed and can only exert tension against septal tensions if they are curved according to the Laplace relation. The bend septal borders, where two septa meet at an angle and are reinforced and held in a tented configuration by a cable, also are curved. We suspect that the curvature of end borders and bend borders are different; cables at ends are probably more strongly curved.

The average lengths of individual border segments of different kinds in dog lung parenchyma have been estimated stereologically. In a lobe inflated to 0.6 V_{L30} (fraction of lobar gas volume of 30 cm H_2O inflating pressure) and having a mean airspace dimension (mean linear intercept, paths through air and one septal thickness) of 130 μm, free edge border segments were 30 μm ± 8 μm (mean ± standard deviation) in length, bend segments, 33 μm ± 13 μm, and junctions, 31 ± 9 μm (5).

The relative numbers of individual border segments may be estimated by dividing the total length of each border type by its individual length. In parenchyma from the same lobe in which average border lengths were estimated, this yields 3,500 end border segments/mm^3, 3,800 bend segments/mm^3, 7,300 junction segments/mm^3 and (assuming a comparable average segmental length) about 1,700 borders of other types/mm^3.

2.2. *Alveolar Septa.*

Each septum is essentially polygonal, with a boundary formed by variously straight or curved border segments of different types. The number of border segments enclosing typical parenchymal septa seems not to be known, but foam septa tend to have five or six sides, on average, and range between 3 to 8 sides (2). The appearance of three-dimensional reconstructions of parenchyma seems consistent with a similar average sidedness.

Are septa flat? Their traces on sections through well-fixed lung appear quite straight, but without careful measurements this is not conclusive proof that the septa themselves are planar. A ruled surface, one that may be generated by the motion of a straight line in three dimensions, can be sectioned to give a straight trace on the section, and not only planes are ruled sections, but also cylinders, cones, saddles, and other quite irregular surfaces too.

An essential fact about parenchyma is that, under static conditions without occluded airways, air pressure must be equal on both sides of every septum. If septa have little resistance to bending (and their composition does not suggest that they have) then their mean curvature (the sum of their two principal, mutually orthogonal curvatures) must be zero. Septa experience at least surface tension and so, by the Laplace relation, if the mean curvature were not zero the septum would exert pressure to one side or another which, by the assumption of uniform gas pressure, cannot happen. Thus, although alveoli themselves may be cup-shaped, the hypothesized requirement for zero mean curvature implies that a septum cannot be. (Closed-cell foam septa (e.g. in suds), however, may be cupped, by virtue of air pressure differences across septa which can arise in this situation.) Surfaces of zero mean curvature are planes or saddles, so septa might be non-planar.

The great majority of septa have at least one cabled border. The fraction of septa bordered only by septal junctions we have estimated to be 7% (5). No septal trace with two free ends has been seen, which suggests that no septum has two free edges. Probably many septa have two or even three cabled borders, although in these cases at least one or two respectively must be bend borders. (Septal traces terminated with a bend trace on each end are common; this is consistent with septa having at least two bend borders.)

2.3 *Interseptal angles, septal borders, and geometric similarity.*

Measurements of the angles between septa which have a common border suggest that geometric similarity is preserved, by and large, but that it may be violated locally. The average of the three interseptal angles around a junction is 120°, by definition (360°/3 = 120°). The standard deviation, however, will be greater than zero if the angles are not all equal to 120°. At 1.0 V_{L30}, we find, the standard deviation is 2.0°, while at 0.4

V_{L30} it is 5.5° (4). (Both figures are averages of values from two lobes.) The simplest implication is that junctions become less regular at low lung volumes and more regular at high. We do not know whether the narrower (or broader) interseptal angles at junctions occur characteristically at a particular orientation to some other structure, such as the alveolar entrance ring.

Interseptal angles also occur along bend borders. The three-dimensional shapes of bends are not known; specifically, it is unclear how curved the bend border is, and how markedly the attached septa depart from planarity. If neither curvature nor non-planarity is severe enough to invalidate our stereological approach, then we may consider unpublished data which indicate that the average bend interseptal angle is 129.5° ± 5.6° at 0.4 V_{L30}, 131.0° ± 5.4° at 0.6 V_{L30}, 136.7° ± 4.6° at 0.8 V_{L30}, and 139.4° ± 2.5° at 1.0 V_{L30}. Where angles change to this extent, geometric similarity cannot be preserved.

Data on bend interseptal angles suggest similarities to the junction interseptal angles: the angles change with lung volume, and dispersion about the mean is narrower at higher volumes. Bends are the only linear parenchymal feature which retain information about the cable/membrane force balance even on two-dimensional sections. (The interseptal angles at junctions in the neighborhood of free edges may be tractable in this respect in three-dimensions, but once sectioned pose apparently insoluble difficulties of interpretation.) If we assume that all cables whether in ends or bends are equally stressed (so that cable connective tissue is at the same elongation ratio throughout), and since we have data suggesting that membrane tension is quasi-uniform, then there would seem to be a definable relationship between observable end cable curvature and cross-section. With bends, the situation could be complicated by the interseptal angle, but as this angle seems to be narrowly distributed, there may be a definable cable curvature-cross section relationship for bend cables, too. Such relationships have not yet been established, however.

As the lung changes volume, septal borders lengthen and shorten proportionate to the cube root of volume, or nearly so (5). The net junction lengths adhere very closely to the cube root relationship, but end border net lengths tend to decrease with deflation faster than this rate, while bend border net lengths tend to decrease more slowly. Whether these differences are due to ends and bends being located in different parts of the duct or at different orientations within it is not known, we believe.

If these changes in net border lengths with volume do reflect the typical behavior of individual borders, than it is probable that most septa nearly maintain geometric similarity during lung volume changes. Septa bounded solely by junction borders seem most likely to maintain similar shapes; however, if such septa are significantly non-planar (as a hexagonal ring may be), then there is a possibility that even they flatten or pucker with volume change and so violate geometric similarity. In the case of the not uncommon septum with both an end border and one or more bend borders, differential lengthening of these two border types causes a shape change. Stretching would be anisotropic in the local tangent plane. If septal tensions themselves are anisotropic in this situation, as may be if tissue tensions contribute to septal tensions (6), then the mean curvature would be non-zero, the radius of curvature being greater (septum flatter) in the direction of greater tension.

3. *Reconciling different microscopical images of parenchyma.*

Lung parenchyma presents different aspects when viewed by different methods, but these various appearances can be reconciled with each other, and in the process our appreciation of parenchymal structure may be improved.

When sectioned thinly and examined by transmission light microscopy (LM), rat parenchymal septa appear as straight line segments meeting at more or less uniform angles (Fig. 4). Alveolar ducts seem recognizable, even if not unambiguously defined, and the free edges of septa seem to project fairly directly toward the duct axis. The image of a rolled-up honeycomb would seem to capture the essence of how we might envision the three-dimensional object of which Fig. 2 is a section. Figure 3 shows tissue from the same animal sectioned by confocal scanning laser microscopy (CSLM), and serial sections of this kind can be reassembled into pseudo-three-dimensional representations (Fig. 4). Here, the dominant impression is of curved surfaces and complexly subdivided outpouchings from the main duct. One sees also many triangular and quadrilateral plate- like structures lining the duct lumen, yet also tipped at an acute angle to the duct's axis, and separated from each other by relatively short lengths of free edges; we dub these "patches".

We decided to explore the nature of these patches further by utilizing several microscopical techniques. Mercer et al. (3) have reported the existence of a "significant lip" overhanging the mouths of cup-shaped alveoli, and the patches do seem to be associated often with lessened alveolar girth, but not always. Being accustomed to the concept of conducting airways and respiratory bronchioles having heavy walls reinforced with plates and bands of connective tissue, one may wonder whether these patches are also strongly reinforced. This would conflict with our understanding of dog lung architecture, however, where only 2.6% of the total septal border length lies along the heavy, alveolated respiratory bronchioles (R borders), which is even less than the percent net length (7.2%) along vessels and non-alveolated airways (F borders). The difference between the percentage of R border length in the dog and the fraction that seems to be associated with patches in the rat lung might be due to interspecific variations in parenchymal constructions (to date, we lack data on septal border lengths in the rat, and cannot decide this question on that basis), but one's informal impression is that both species are more similar in their fractions of respiratory bronchioles than different.

The patches seem to be as thin as typical septa, not thick like bronchiolar walls. They seem in some measure more densely reinforced with connective tissue than other septa, but also to be supplied with capillaries. These points may be demonstrated in two ways: first, by examining both the front and back sides of patches and, second, by examining two-dimensional sections through selected patches.

(These illustrations are anecdotal and not supported by statistically significant numbers of observations; they exemplify, however, an approach to utilizing varied microscopical techniques to explore alveolar and ductal architecture.)

Fig. 5 shows the upper left quarter of Fig. 4A from front and back, and two arrows point out 2 patches which seemingly could be thick and blocky, but which turn out to be deeply hollowed out from the back side.

Fig. 6 shows the appropriate CSLM section superimposed upon the reconstruction, and in two locations, indicated by arrows, the bright areas of the section which indicate high connective tissue concentrations fade out towards the centers of the patches. A similar configuration chosen from the same tissue and viewed by conventional LM at high magnification (Fig. 7) shows a gap between two collagen and/or elastin-containing cables bordering the sectioned patch; the patch tissue is also clearly supplied with capillaries on both sides.

It will continue to be fascinating to understand the structure of normal, mature parenchyma, to trace the routes by which it emerges during development and becomes disrupted with disease, and to comprehend how function and dysfunction are influenced by parenchymal architecture.

ACKNOWLEDGEMENTS

This work has been supported by National Heart, Lung, and Blood Institute Grants HL-26863 and HL-33009.

REFERENCES

1. Carson J. On the elasticity of the lungs. Philosophical Transactions Roy. Soc. London Part 1:29-44, 1820.

2. Matzke EB. The three-dimensional shapes of bubbles in foams. *Proc. Nat. Acad. Sci.* USA 31:281-289, 1945.

3. Mercer RR, JM Laco, JD Crapo. Three-dimensional reconstruction of alveoli in the rat lung for pressure-volume relationships. *J. Appl. Physiol.* 62:1480-1487, 1987.

4. Oldmixon EH, JP Butler, FG Hoppin, Jr. Dihedral angles between alveolar septa. *J. Appl. Physiol.* 64(1):299-307, 1988.

5. Oldmixon EH, JP Butler, FG Hoppin, Jr. Lengths and topology of alveolar septal borders. *J. Appl. Physiol.* 67(5):1930-1940, 1989.

6. Oldmixon EH, FG Hoppin. Distribution of collagen and elastin in canine lung alveolar parenchyma. *J. Appl. Physiol.* 67(5):1941-1949, 1989.

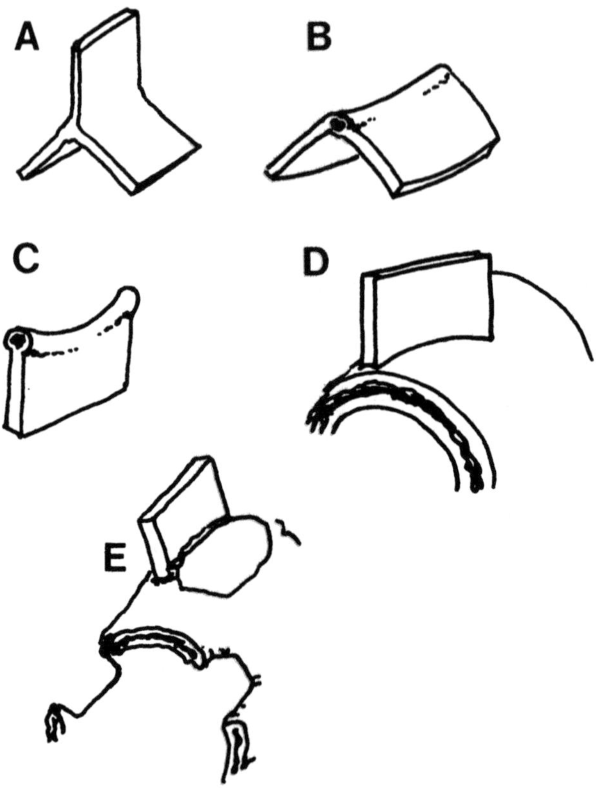

Fig. 1 The five types of septal borders seen in dog lung parenchyma, pictured in order of decreasing net length. A. Junction: 3 septa meet along a line; no additional connective tissue involvement. B. Bend: 2 septa attached to a connective tissue cable (see text). C. End (or free edge): 1 septum attached to a cable. D. Fixed attachment: 1 septum abutting a non-alveolated airway, or vessel (venule, arteriole, or larger). E. Respiratory bronchioles attachment: 1 septum attached to relatively heavy luminal bands of alveolated bronchiole.

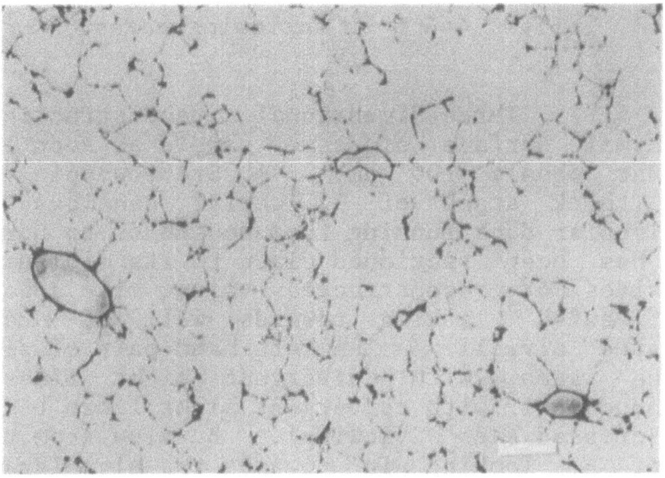

Fig. 2 Light micrograph of normal rat lung parenchyma; tissue from lungs (rat #52) held at 0.4 V_{L30}, prepared by vascular perfusion fixation/dehydration, sectioned 0.4 μm thick, stained with Phloxine B-methylene blue-Azure II. Bar equals 100 μm.

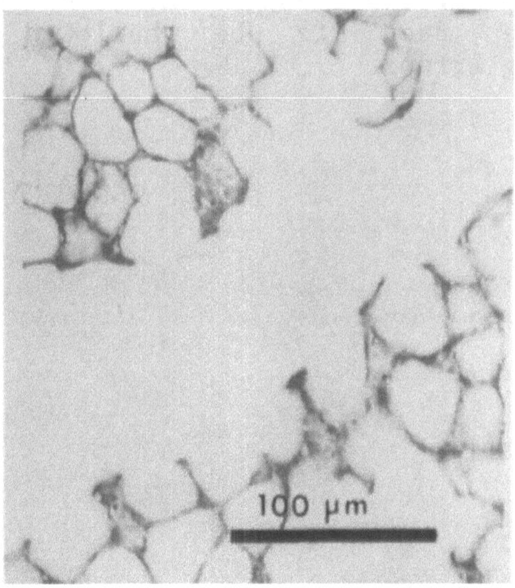

Fig. 3 Confocal scanning laser micrograph of same sample as Fig. 4. Image shown is a mosaic of 4 smaller images taken with 40X objective, 1.25X auxiliary magnification. Pale grey surrounding tissue is artifactual. Tissue stained with Lucifer yellow CH. Bar equals 100 μm.

(Figs. 4 & 5 on following pages.)

Fig. 4 Three-dimensional reconstruction of parenchymal surface of same sample as seen in Fig. 3. Bar equals 100 μm. A. Stereoscopic view of 96 μm thick stack of 32 serial sections. Note the alveolar duct running from southeast to northwest which has been sectioned close to its midline. The near sides of reconstructed volume lie close to ductal axis, looking towards wall of duct and associated alveoli. The left-hand pair of pictures may be seen with divergent sight lines, the right-hand pair with convergent sight lines (that is, with "crossed-eyes" viewing). B. View from the far side of A, looking in towards the alveolar ductal axis.

Fig. 5 Portion of Fig. 4 data reconstructed at higher resolution to show two "patches" (see text) from two opposite directions. Bar equals 100 μm. A. View from ductal axis. Two patches are indicated by arrows 1 and 2. B. From opposite side, patches are seen to be hollowed out; see corresponding arrows.

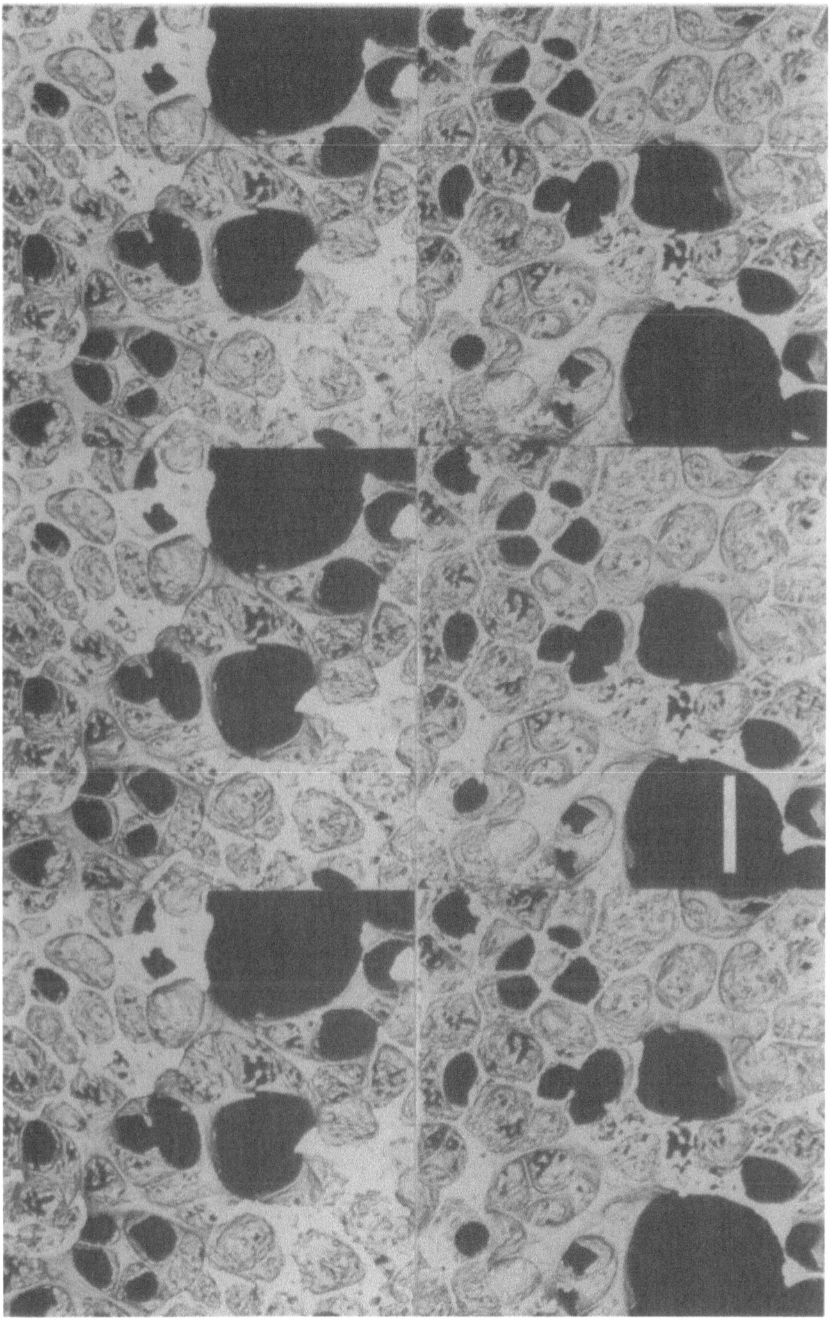

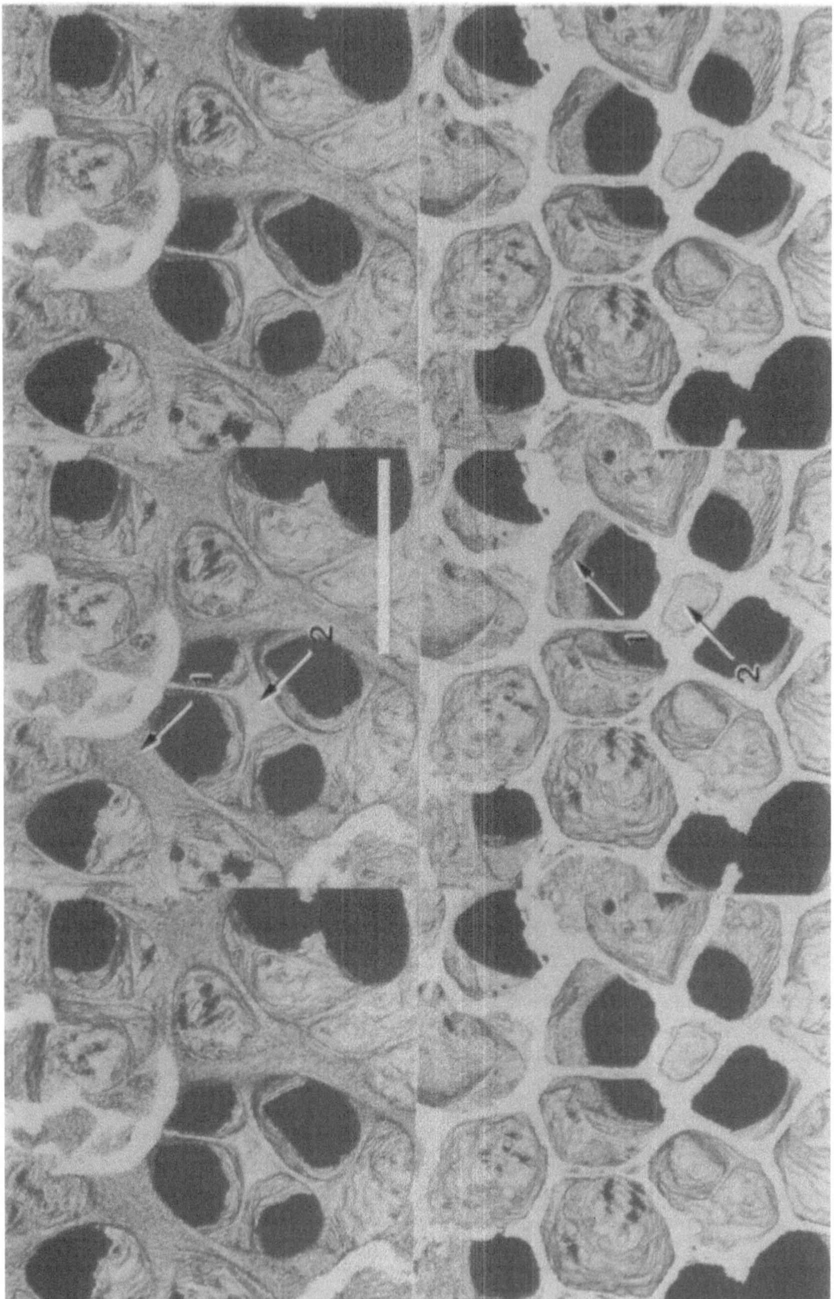

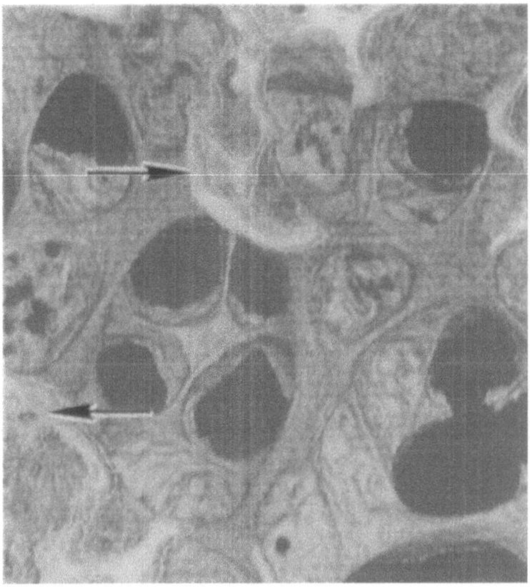

Fig. 6 CSLM section superimposed upon three-dimensional reconstruction, showing diminished fluorescence (arrows) in center of patch indicating that lesser amounts of connective tissue occur in these areas.

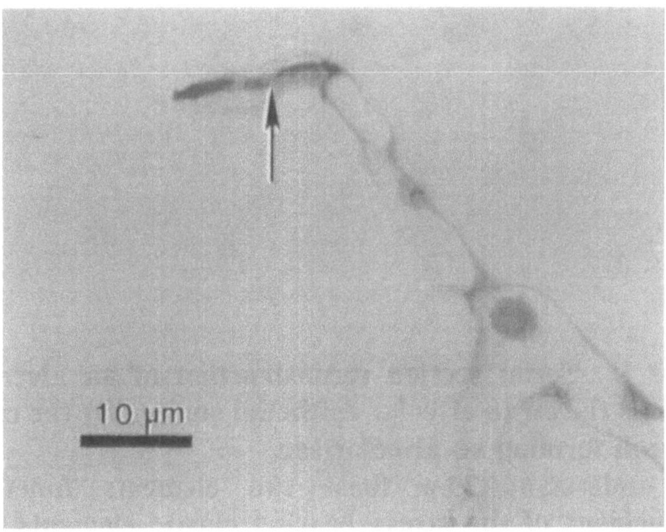

Fig. 7 LM section showing diminished density (indicative of collagen or elastin) in gap (arrow) between cables bordering septum which probably would appear as a patch if viewed in three dimensions. Bar equals 10 μm.

VOLUME-PRESSURE HYSTERESIS OF THE LUNGS
ROBERT R. MERCER, DUKE UNIVERSITY

Recoil due to tissue elastic forces and surface tension are the principal mechanisms which counter balance the inflating transpulmonary pressure of the lungs. The stress-strain properties of the elastic elements and surface tension-area relationships of surfactant are not sufficient to describe the mechanical behavior of the lungs because of the complex arrangement of the alveoli and alveolar ducts (Figure 1.).

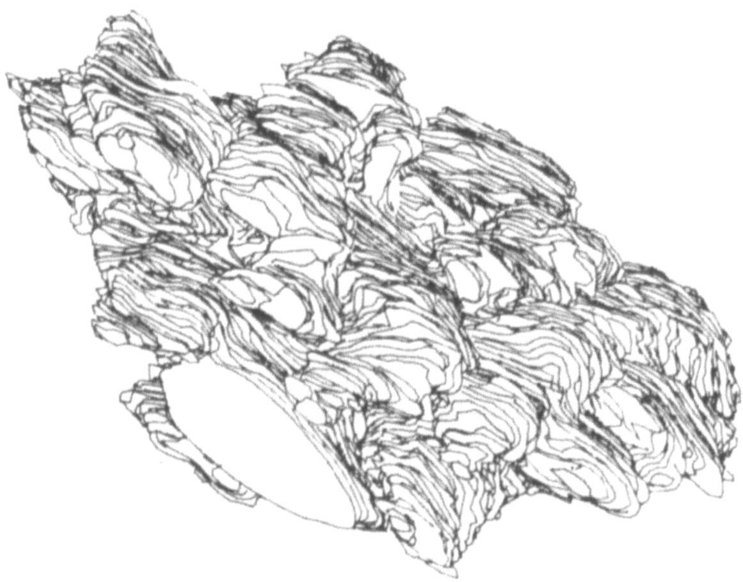

Figure 1. Serial section reconstruction of an alveolar sac. Shown is the air to alveolar epithelial surface for the collection of alveoli forming an alveolar sac.

To understand how these two elements function the arrangement of the stress bearing elastic elements and the radius of curvature at the air to tissue interface are necessary. Such measurements may be obtained from serial section reconstructions (ex. Figure 1).

For instance, a central question in lung micromechanics is

whether all alveoli participate throughout the entire volume-pressure curve. Alternately, sequential recruitment-derecruitment of alveoli might arise from surface tension effects on a population of alveoli of differing size.

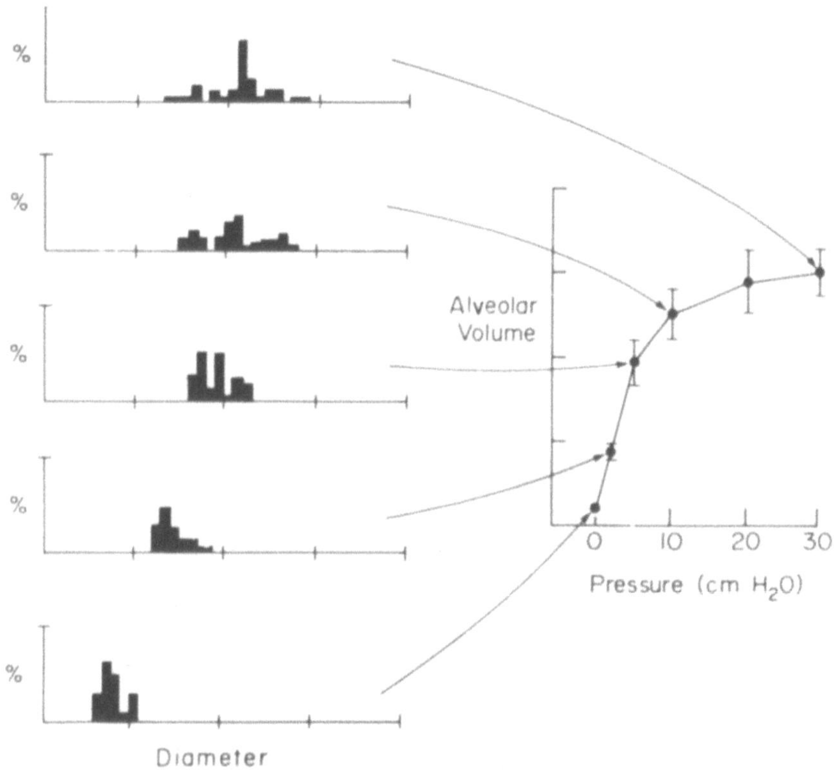

Figure 2. Histograms of alveolar diameter during lung deflation.

Only recruited alveoli would increase their size during lung inflation. This can be evaluated by determining alveolar diameter at different stages of lung inflation (Mercer and Crapo, 1988). The histograms of alveolar diameter in Figure 2 indicate that as the lungs deflate, the entire population of alveoli decrease in size. This and other measurements at different levels of lung inflation (Mercer et al, 1987) indicate that all alveoli in a normal lung participate in volume changes throughout the whole lung pressure-volume curve.

From similar 3D based analysis the spatial distribution of connective tissue fibers, Figure 3, has been determined in different species (Mercer and Crapo, 1990, Mercer et al, 1990).

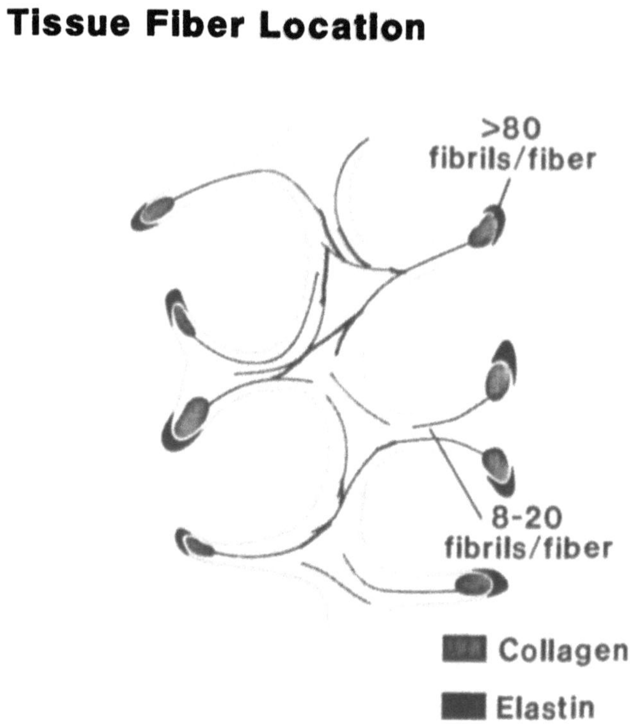

Figure 3. Schematic representation of the spatial distribution of collagen and elastin fibers. Collagen and elastin fibers are a large fraction of the alveolar tissue present in the edge of the alveolar septa.

The results indicate a comparable distribution of connective tissue fibers relative to alveolar-alveolar duct structure in different species. As illustrated, elastin fibers are in close association with the collagen fibers in the edge of the alveolar septa forming the alveolar duct wall. In addition to the close proximity to the collagen fiber, serial section analysis demonstrates that collagen fibrils from adjacent collagen fibers form a mechanical linkage between elastin and collagen fibers (Figure 4).

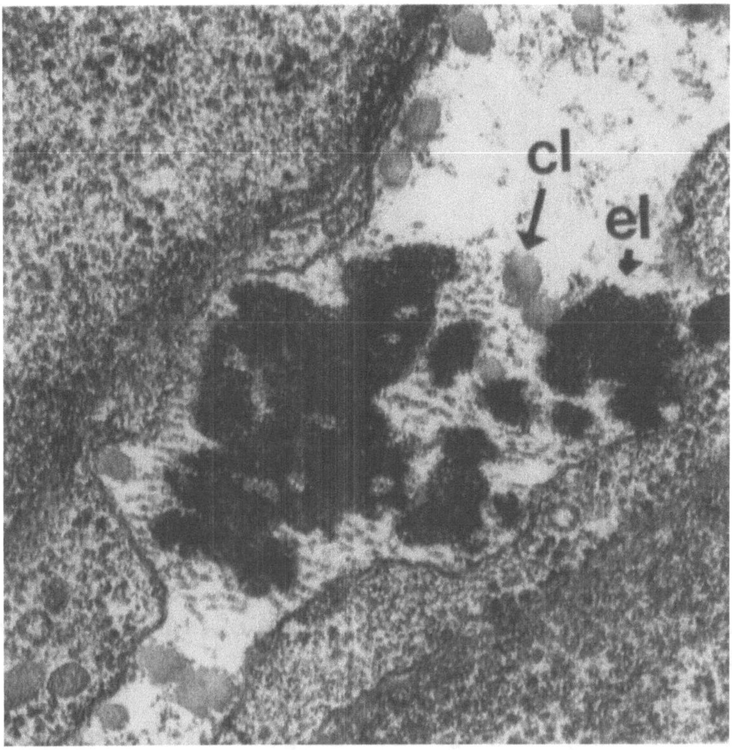

Figure 4. Electron micrograph of an elastin fiber with intermingled collagen fibrils.

Given the dramatic differences in stress-strain properties from in-vivo analysis of collagen and elastin rich tissue, how can we resolve the close spatial proximity and mechanical linkage between these two dissimilar connective tissue fibers? To answer this question we have studied the structure of collagen fibrils at different levels of lung inflation. Figure 5 shows a reconstruction of 4 collagen fibrils from a collagen fibers in a lung preserved near functional residual capacity. More extensive 3D analysis (Mercer and Crapo, 1990) indicates that collagen fibrils from straight segments of collagen fibers have a wave-like or zig-zag arrangement. The stress-strain properties of collagen fibers at low levels of lung inflation may be more distensible than previously thought due to this wave-like arrangement of collagen fibrils forming the collagen fibers.

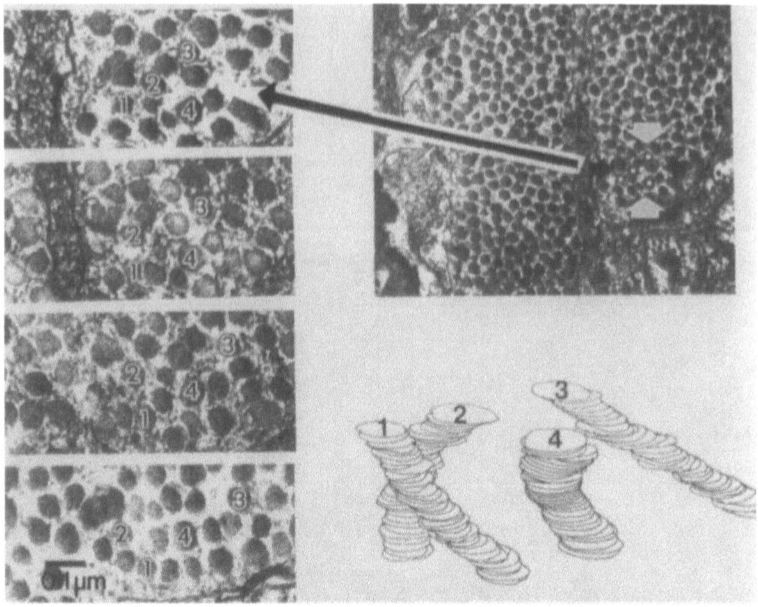

Figure 5. Serial section reconstruction of individual collagen fibrils.

REFERENCES

Mercer, R. R., and J. D. Crapo. Three-dimensional reconstruction of the rat acinus. J. Appl. Physiol. 63: 785-794, 1987.

Mercer, R. R., and J. D. Crapo. The structure of the gas exchange region of the lungs determined by three-dimensional reconstruction. In: Toxicology of the Lung, D. E. Gardner, J. D. Crapo, and E. J. Massaro, eds., Raven Press, New York, 1988.

Mercer, R.R., and J.D. Crapo. The spatial distribution of collagen and elastin fibers in the lungs. J. Appl. Physiol. (in press), 1990.

Mercer, R.R., M.L. Russell and J.D. Crapo. Species variations in alveolar septal wall connective tissue fibers. Am Rev Resp Dis, 141,A714, 1990.

LUNG TISSUE MECHANICS. Stephen J. Lai-Fook, University of Kentucky

Introduction

Lung parenchyma consists of millions of interconnecting cellular units, called alveoli, that are homogeneously distributed between two parallel repeatedly branching networks, the pulmonary airways and vasculature. On the alveolar scale (100 μm) the forces in the tissue membranes are heterogeneously oriented. However on a scale that encompasses several alveoli, the macroscopic properties of lung parenchyma can be defined in terms of average stesses and average strains. On this scale the lung parenchyma is fairly homogeneous and isotropic (1). For small quasistatic deformations, the lung parenchyma is assumed to behave elastically. However, the lung is known to exhibit both viscous and plastic properties. The lung also undergoes changes in volume that are outside the range of linear elasticity. For large changes in volume, the inelastic behavior of the lung is accentuated. Thus, the concept of elasticity as applied to the lung is restricted to small changes in distortion.

The focus of this note is to describe the properties of lung parenchyma as an elastic solid continuum within the mathematical framework of infinitesimal elasticity. In this approach the lung is considered to be in a state of uniform inflation at a fixed inflation (transpulmonary) pressure, and small distortions are imposed on this isotropic state. The solution of nonuniform deformation problems requires the knowledge of only two independent elastic moduli, such as the bulk modulus and shear modulus. However, the elastic constants are functions of the inflation pressure (prestress) and have to be measured at each inflation pressure. Another approach to solving lung deformation problems considers distortions from the unstressed state. However, this approach as it applies to the lung requires a large deformation description for the material behavior and more complex analysis.

Linear Elastic Constants

Quasistatic deformation experiments have been used to measure two linear elastic constants that are sufficient for solving linear elasticity problems (8, 15, 17). One of these constants, the bulk modulus (K), is easily measured using the pressure-volume behavior of isolated lungs. The bulk modulus is calculated using the slope ($\Delta P/\Delta V$) of a small pressure-volume loop imposed on any lung volume (V): $K = V(\Delta P/\Delta V)$. The bulk modulus is a unique function of transpulmonary pressure (P) and independent of lung volume history. The bulk modulus increases exponentially with P; the values for K range from 4P to 6P for P in the range 4 to 25 cmH_2O. Another elastic constant, the shear modulus, is obtained from indentation tests on inflated lungs. The shear modulus (G) is proportional to transpulmonary pressure and has a value of 0.7P. Other elastic constants can be computed from the values of K and G. The Young's modulus has a value of 2P and Poisson's ratio lies between 0.37 and 0.45. Note that the shear modulus, a measure of the lung parenchyma to deform without a change in volume, is about 7 times smaller than the bulk modulus. This implies that in many situations the lung is more easily deformed by a change in shape rather than by a change in volume.

Determinants of Elastic Constants

A description of the elastic constants in terms of the transpulmonary pressure results in two important consequences. First, the elastic constants defined in terms of transpulmonary pressure are fairly invariant among different species such as, dog, pig, horse and rabbit. (8, 14, 15, 19, 21). Second, the effect of age is to increase the value of the bulk and shear moduli at any transpulmonary pressure; that is, the lung parenchyma becomes stiffer with age (19). This effect of age is probably due to deterioration of elastin in lung parenchyma.

Both tissue forces and surface forces contribute to the elastic constants of lung parenchyma. Surface forces arise from the alveolar air-liquid interface and its properties are attributed to pulmonary surfactant. The relationship between the tension in the walls of the microstructure to the macrostrucure properties has been studied extensively (3, 13). Tension in the alveolar walls is the major factor that contributes to the bulk and shear moduli behavior with increasing transpulmonary pressure. An increase in alveolar surface tension is thought to decrease the bulk modulus and increase the shear modulus (21). The contribution of the surface forces that result from a foam-like structure accounts for about half of the resistance to shear (22). Indentation tests show that in liquid-filled lungs the shear modulus has the same relationship to P as in the air-filled lung. Thus changes in alveolar configuration due to surface forces may also be important.

Application to Static Nonuniform Deformation Problems

The importance of the linear description of elasticity to the lung lies in its particular application to the solution of problems in pulmonary physiology. We shall describe three problems that have received attention.

Interaction among blood vessels, airways and lung parenchyma. The interstitium of the lung surrounds blood vessels and airways so that the perivascular and peribronchial pressures are important to understanding the mechanisms responsible for fluid exchange and lung fluid balance. Lung inflation has been shown to increase the volume of large intraparenchymal blood vessels. A continuum analysis has been used to study the interaction between the blood vessel and the surrounding lung parenchyma (14, 15). The vessel is considered as a thin-walled elastic tube embedded in a cylindrical hole in an elastic continuum, the lung parenchyma. The diameter of the vessel is determined by its transmural pressure, the difference between the vascular pressure and the perivascular pressure. The deformation of the parenchyma surrounding the vessel is given by Lame's solution for the expansion of a hole in an elastic continuum. Deformation of the hole in the parenchyma is assumed to occur from reference states of uniform inflation that are given by the pressure-volume behavior of the lung. The pressure acting on the hole boundary at the reference states is the pleural pressure (20). Measurements of the diameter of intraparenchymal arteries and veins as a function of vascular pressure at different transpulmonary pressures were reconciled with the elastic constants of the lung parenchyma. The perivascular pressure, the pressure exerted by the lung parenchyma on the vessel, was obtained from the analysis. The analysis showed that perivascular pressure was near the pleural pressure at functional residual capacity and is reduced below pleural pressure with lung inflation. A reduction in vascular pressure reduces perivascular pressure. A similar analysis of bronchial (airway) pressure-diameter behavior showed that peribronchial pressure remained near the pleural pressure with lung inflation. These two studies suggest that during pulmonary edema formation the periarterial interstitium is the perferential site of edema accumulation.

Effect of the airway on arterial distortion. The proximity of the airway to the artery results in an asymmetrical distortion of these cylindrical structures (16). A finite element analysis of the interaction among the bronchus, artery and the lung parenchyma showed that with a reduction of vascular pressure, the bronchus is pulled into an oval shape with no change in the mean peribronchial pressure. This bronchial shape change is accompanied by a reduction in vascular diameter and an oval arterial cross-sectional geometry. The stress analysis shows a stress concentration present in the adjoining region between the artery and bronchus. This site will be the focal point of emphysematous air dissection and tearing during lung hyperinflation and the preferential site for fluid collection in the inital stages of pulmonary edema formation.

Effect of gravity on regional lung expansion. The effect of gravity acting on the upright lung and heart on the distribution of lung volume has been studied using linear elasticity (2). The analysis showed that the vertical gradient in transpulmonary pressure increased from 0.2 cmH_2O/cm height without the weight of the heart to 0.3 cmH_2O/cm with the weight of the heart. Increasing heart weight by filling it with mercury increased the gradient to 0.5 cmH_2O/cm (10). Simulation of lung inflation showed that the vertical gradient increases at high lung volumes (2). These results confirm and extend the pioneering work of West and Matthew who used a nonlinear stress-strain law in their finite element analysis (23). The weight of the heart has also been shown to be important in explaining the larger vertical transpulmonary pressure gradient in animals in the supine position than in the prone position (24). In the supine position the heart moves downwards with gravity thereby compressing the lower lung regions and expanding the upper regions. These effects are accentuated by a diaphragm that is much more compliant than the rib cage (6). In the prone position the heart is supported by a rigid sternum and gravity has little effect on its position.

Stress Waves in Lung Parenchyma

The foregoing studies show good agreement between quasistatic deformation experiments and predictions using linear elasticity. The description of the lung as a linear elastic continuum also would predict the existence of small amplitude stress waves. The transmission properties of the lung are important to understanding how the lung responds to impact loads as would occur during automobile accidents and during combat (5). The velocity of longitudinal waves

(v_l) is related to the elastic constants and lung density (ρ) by the following equation

$$v_l = [(K + 4G/3)/\rho]^{1/2}$$

The velocity of shear waves (v_s) is

$$v_s = (G/\rho)^{1/2}$$

Recently, in experiments performed in isolated sheep, dog and horse lungs, the velocities of small amplitude stress waves have been shown to correlate well with measured elastic constants (4, 11, 12, 18). Longitudinal and shear waves velocities measured from signals generated on the surface of isolated lungs showed that both wave velocities increase with transpulmonary pressure in accord with theoretical predictions (11). Longitudinal and shear wave velocities were in the range 300-600 cm/s and 130-250 cm/s, respectively, for transpulmonary pressures of 5-15 cmH_2O. Increases in lung density due to increases in vascular volume and edema formation resulted in a reduction in lung density as dictated by theory (12). These waves are transmitted at relatively low frequency (<70Hz) and long wave length (2-10 cm). An elasticity analysis of the lung parenchma covered by a taut pleural membrane predicted the existence of a cutoff frequency above which no surface waves are propagated (18). These results were confirmed in experiments using horse lungs.

In summary, the theory of linear elasticity has proved useful in describing many lung deformation problems. Recent studies also show that linear elasticity is able to describe the propagation of small amplitude stress waves through the lung parenchyma.

Acknowledgements.

This research was supported by research grants HL-40362 and HL-36597 from National Heart, Lung and Blood Institute.

References

1. Ardila R, Horie T, Hildebrandt J. Macroscopic isotropy of lung expansion. *Respir. Physiol.* 20: 105-115, 1974.
2. Bar-Yishay E, Hyatt RE, Rodarte JR. Effect of heart weight on distribution of lung surface pressures in dogs. *J. Appl. Physiol.* 61:712-718, 1986.
3. Budiansky B, Kimmel E. Elastic moduli of lungs. *J. Appl. Mech.* 54: 351-358, 1990.
4. Butler JP, Lear JL, Drazen LM. Longitudinal elastic wave propagation in pulmonary parenchyma. *J. Appl. Physiol.* 62: 1349-1355, 1987.
5. Fung YC, Yen RT, Tao ZL, et al. A hypothesis on the mechanism of trauma of lung tissue subjected to impact load. *J. Biomech. Eng.* 110:50-

56, 1988.
6. Ganesan S. Lai-Fook SJ. Finite element analysis of regional lung expansion in prone and supine positions: effect of heart weight and diaphragmatic compliance. *Physiologist.* 32:191, 1989.
7. Goshy M, Lai- Fook SJ, Hyatt RE. Perivascular pressure measurements by wick catheter technique in isolated dog lobes. *J. Appl. Physiol.* 46: 950-955, 1979.
8. Hajji MA, Wilson TA, Lai-Fook, SJ. Improved measurements of shear modulus and pleural membrane tension of the lung. *J. Appl. Physiol.* 47: 175-181, 1979.
9. Hoffman EA, Ritman EL. Effect of body orientation on regional lung expansion in dog and sloth. *J Appl Physiol.* 59:481-491, 1985.
10. Hyatt RE, Bar-Yishay E, Abel MD. Influence of the heart on the vertical gradient of transpulmonary pressure in dogs. *J. Appl. Physiol.* 58: 52-57, 1985.
11. Jahed M, Lai-Fook SJ, Bhagat PK, Kraman SS. Propagation of stress waves in inflated sheep lungs. *J. Appl. Physiol.* 66:2675-2680, 1989.
12. Jahed M, Lai-Fook SJ, Bhagat PK. The effect of vascular volume and edema on stress wave propagation in inflated lungs. *J. Appl. Physiol.* In press.
13. Kimmel E, Kamm RD, Shapiro AS. A cellular model of lung elasticity. *J. Biomech. Eng.* 109:126-131, 1987.
14. Lai-Fook SJ. A continuum mechanics analysis of pulmonary vascular interdependence in isolated dog lobes. *J. Appl. Physiol.* 46: 419-429, 1979.
15. Lai-Fook SJ. The elastic constants of lung parenchyma: the effect of pressure-volume hysteresis on the behavior of blood vessels. *J. Biomech.* 12:757-764, 1979.
16. Lai-Fook SJ, Kallok MJ. Bronchial-arterial interdependence in isolated dog lung. *J. Appl. Physiol.* 52:1000-1007, 1982.
17. Lai-Fook SJ, Wilson TA, Hyatt RE, Rodarte JR. Elastic constants of inflated lobes of dog lungs. *J. Appl. Physiol.* 40: 508-513, 1976.
18. Man C-S, Jahed M, Bhagat PK, Lai-Fook SJ. Effect of pleural membrane on the propagation of Rayleigh-type surface waves in inflated horse lungs. *Physiologist.* 32:191, 1989.
19. Mansell AL, Moalli RR, Calista CL et al. Elastic moduli of lungs during postnatal development in the piglet. *J. Appl. Physiol.* 67: 1422-1427, 1989.
20. Mead J, Takishima T, Leith D. Stress distribution in lungs: a model of pulmonary elasticity. *J. Appl. Physiol.* 28: 596-608, 1970.
21. Stamenovic D, Smith JC. Surface forces in lungs. III. Alveolar surface tension and elastic properties of lung parenchyma. *J. Appl. Physiol.* 60:1358-1362, 1986.
22. Stamenovic D and Wilson TA. The shear modulus of liquid foam. *J. Appl. Mech.* 51:229-231, 1984.
23. West JB, Matthew FL. Stresses, strains and surface pressures in the lung caused by its weight. *J. Appl. Physiol.* 32: 332-345, 1972.
24. Wilson TA, Liu S, Marguiles SS. Deformation of the dog lung in the chest wall. *Faseb J.* 3: A240, 1989.

Pleural Mechanics

James R. Ligas, University of Connecticut Health Center, Farmington CT 06032

The visceral pleura is a thin membrane which completely surrounds the lung and is tightly adherent to the parenchymal tissue. It contains elastic fibers which form an irregular network with no preferential orientation [1]. Although many physiologic roles have been proposed for this membrane, we limit our focus to the effects of the pleura on lung mechanics. Specifically, what is the relationship between stress and strain for the pleural membrane? And what implication does this relationship have for the overall mechanical function of the lung?

Constitutive equations for pleural tissue:

Hildebrandt et al. [2] studied the mechanical properties of various soft tissues, including pleura. Uniaxial loading of excised pleural strips, together with pressure loading of circular tissue sheets, revealed that there was a nonlinear relationship between stress and strain similar to that of other soft tissues. Some hysteresis was observed as well. The investigators concluded that the pleural membrane was probably incompressible and isotropic within the plane of the membrane. Furthermore, a simple polynomial equation relating stress to strain fit the data reasonably well.

Hajji et al. [3] were next to report on the relationship between pleural membrane tension versus change in membrane area during inflation of a lobe of canine lung. After expansion to a given volume, a cylinder of known weight was allowed to rest upon and indent the pleural surface. An analysis for an elastic half space bounded by a membrane, together with determination of the shear modulus for the underlying parenchymal tissue, allowed an indirect estimation of pleural membrane tension as a function of area change. Again, this relationship was nonlinear. The investigators also noted that pleural tension increased markedly near full lung expansion:

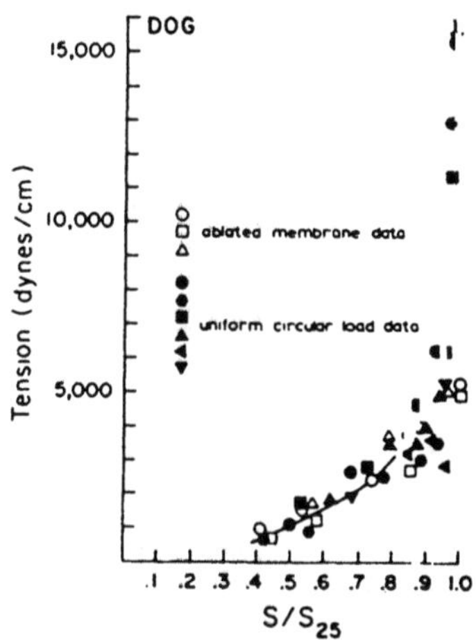

Membrane tension versus area ratio where S25 is lobe surface area at p = 25 cm H2O. From [3].

FIGURE 1

Direct assessments of pleural tension as a function of pleural area change came from others. Ligas et al. [4] measured the tension-elongation ratio behavior of excised canine pleural tissue from circular pleural specimens loaded with an applied pressure (Figure 2).

These values agreed closely with those of

Hajji et al. when suitable assumptions were made about the relationship between area ratio and pleural stretch [4].

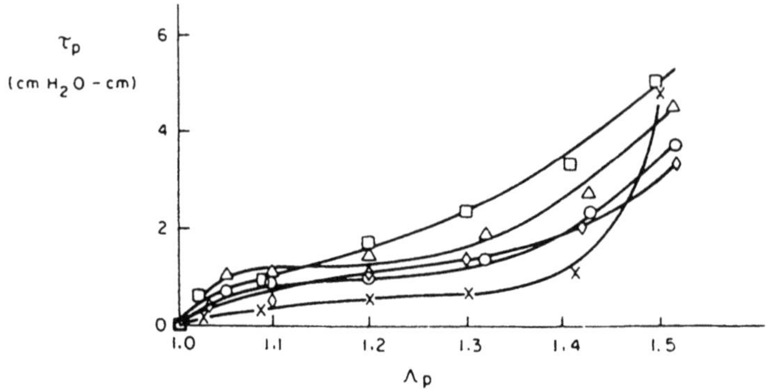

Relationship between pleural tension and extension ratio for 5 specimens. From [4].

FIGURE 2

Analytically, improvements in the one-dimensional constitutive relationship proposed for soft tissues in general came originally from Wilson [5] and were extended by Stamenovic [6]. The two-dimensional network of randomly oriented fibers noted on microscopy was coupled with the exponential relationship between length and tension to write an energy density function for the pleural membrane. The proposed constitutive equation fit the data reasonably well. Furthermore, these investigators punched circular holes in the pleural membrane and proceeded to stretch the specimens. The deformation field about the holes compared well with the theoretical predictions.

Finally, Humphrey and co-workers [7,8] proposed two- and three-dimensional pseudostrain-energy functions for the pleural membrane. They then excised specimens of canine visceral pleura and subjected 3.5 cm squares to biaxial testing according to three protocols. In the first, uniform forces were applied in both directions. In the second, the force in the x-direction was kept constant while the membrane was stretched in the y-direction. And finally, the length of the specimen in the x-direction was maintained constant while the material was stretched in the y-direction. During these experiments, the rate of strain was varied 100-fold. Once again, the relationship between tension and strain was found to be nonlinear, and was also found to be relatively rate-insensitive. The material appeared to be incompressible and isotropic within the plane of the membrane. Some hysteresis was present as well (Figure 3).

These investigations show that an approximately exponential stress-strain relationship fits the pleural data fairly well. However, the assumption of elasticity, or pseudoelasticity, is only an approximation to the mechanical behavior of the membrane inasmuch as there is some degree of hysteresis observed by all experimenters.

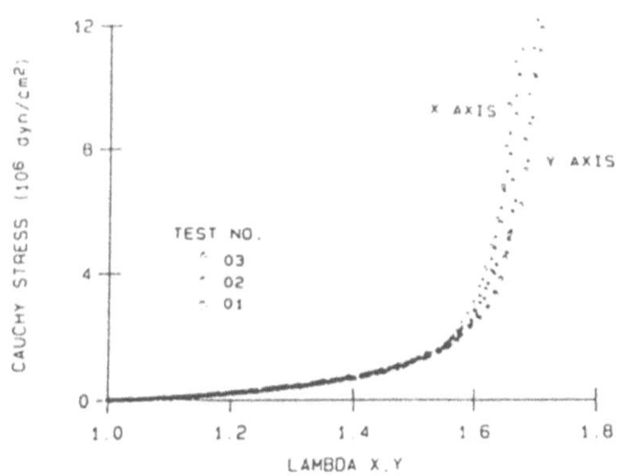

Cauchy stress vs. stretch ratio for both axes.
Data are from three loading uniform stretch tests.
From [7].

FIGURE 3

Implications for lung mechanics:

In the larger picture, however, what most interests mechanicists is the effect of the pleura on the expansion of the underlying lung parenchyma. To assess this influence, Nagao [9] compared the pressure-volume relationship of an excised canine lobe to that of the pleural sack made by stripping the pleura free of the underlying lung tissue. During deflation, at equivalent volumes the pressures required to inflate the pleural sack were much less than those required to inflate the intact lobe until volume was reduced to approximately 50% of total lung capacity. At that point the pressure required for pleural sack inflation was roughly 15-20% of that required to inflate the lobe to that same volume. He concluded that at low lung volumes the pleura bore a significant portion of the pressure load. One technical objection to this comparison is that the pleural sack, stripped free of the underlying parenchyma, need not maintain the same shape as it has when attached to the lung tissue. But a more serious objection to this comparison is that one must know the exact structural relationship between the pleural membrane and the outermost alveoli to assess the effects of the pleura on underlying parenchymal expansion [4]. Two of many possible structures are shown in Figure 4.
With the first structure, the pleural tension does not affect lung expansion. The pressure difference across the pleural surface must be equal to the parenchymal stress, and the pleural tension is merely the mode of transmission. In the second model, the parenchymal stress at the lung surface is equal to the pressure difference minus a component dependent upon the pleural membrane tension and the local radii of curvature:

$$\Sigma_s = (P_i - P_o) - \tau_p(1/R_1 + 1/R_2)$$

From the data presented above, one can infer that if the structure is similar to that in the second figure, pleural tension can reduce the applied surface stresses transmitted to the underlying parenchyma in regions of high surface curvature [4,8]. In addition, those studies which have been carried out to high strains [7] and the constitutive relationships themselves [6] suggest that at very large strains the pleural membrane tension may increase quickly and thereby limit lung expansion.

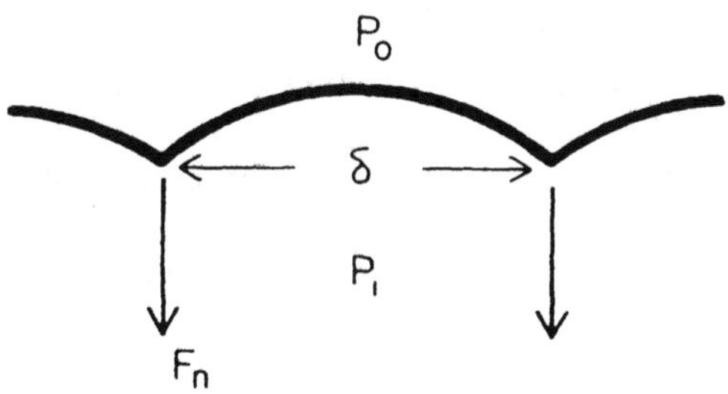

"Cobblestone pleura" model. Fn is the normal alveolar wall force, P represents the pressures internal and external to the pleura, and δ the alveolar diameter at the pleural surface. From [4].

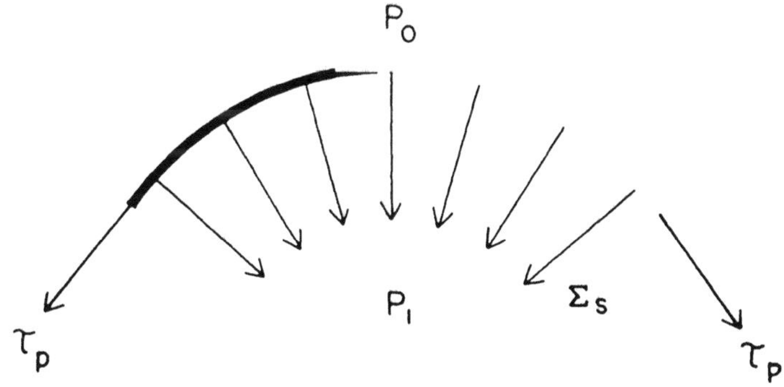

An alternative pleural model. Again P represents the internal and external pressures. Σs is the parenchymal surface stress, τp the pleural tension. From [4].

FIGURE 4

Conclusion:

It is important to note that all the above data were obtained from in vitro experiments. In vivo data are needed to assess whether excision of the pleura alters its mechanical properties. In addition, although much data and several proposed constitutive equations for pleural membrane exist, one needs to know how the mechanical properties specified in these formulations relate to microstructural elements of the pleural sheet. In addition, all of the above formulations are concerned with elasticity or pseudoelasticity. By the very nature of the experiments and analyses viscoelastic or plastic effects are excluded.
Finally, to assess the effects of the pleura in situ on the underlying lung parenchyma, the microstructure of the parenchymal/pleural interactions must be elucidated and the stresses applied to the visceral pleura by the intrapleural space better understood.

References:

1. von Hayek H: The Human Lung. New York: Hafner, 1960.

2. Hildebrandt J, Fukaya H, Martin CJ: Simple uniaxial and uniform biaxial deformation of nearly isotropic incompressible tissues. Biophysical J. 9:781-791, 1969.

3. Hajji MA, Wilson TA, Lai-Fook SJ: Improved measurements of shear modulus and pleural membrane tension of the lung. J. Appl. Physiol. 47:175-181, 1979.

4. Ligas JR, Primiano FP Jr, Saidel GM: Static mechanics of excised whole lung: pleural mechanics. Ann. Biomed. Engr. 12:437-448, 1984.

5. Wilson TA: Mechanics of the pressure-volume curve of the lung. Ann. Biomed. Engr. 9:439-449, 1981.

6. Stamenovic D: Mechanical properties of pleural membrane. J. Appl. Physiol. 57:1189-1194, 1984.

7. Humphrey JD, Vawter DL, Vito RP: Mechanical behavior of excised canine visceral pleura. Ann. Biomed. Engr. 14:451-466, 1986.

8. Humphrey JD: A possible role of the pleura in lung mechanics. J. Biomech. 20:773-777, 1987.

9. Nagao K: Experimental studies on mechanical properties of the visceral pleura. Nihon Univ. J. Med. 15:307-327, 1973.

RESPIRATORY FLUID MECHANICS AND TRANSPORT

RESPIRATORY FLUID MECHANICS AND TRANSPORT

Mary A. Farrell Epstein

The last decade was a period of significant progress in our understanding of gas phase fluid mechanics and species or mass transport in the airways. Several factors contributed to this advance. Detailed measurements of airway geometry from the nasal passages to the alveolar level, made possible by advances in computer-based imaging and microscopy systems, revealed the complex geometry in which gas flow and mass transport take place, even in normal lungs. Introduction of high-frequency ventilation (HFV) techniques with frequencies of 20 or more times resting breathing rates underscored the sensitivity of pressure gradients and velocity distributions to subtle changes in airway geometry and mechanical properties, and focused attention on mechanisms for mass transport under these conditions. The need to understand factors affecting gas exchange at elevated frequencies reopened questions of the interaction of respiratory mechanics with gas exchange. A third factor was the ready availability of computational resources through introduction of the microcomputer, and the continued enhancement of execution speed, memory and storage in medium and large-scale computer systems. These permitted extensive investigation of nonlinear parameter estimation techniques and complex computer simulations as innovative approaches for data analysis.

One important characteristic of the research during this decade was the increasing use of analytical approaches, drawn from physics, mathematics and engineering, to airway fluid mechanics research. This section reflects the continued development of new methodologies and recent progress on key problems of gas flow and mass transport in the airways.

Forced oscillation techniques are used frequently for characterizing respiratory mechanics because these methods permit measurement during spontaneous or assisted breathing and without the active cooperation of the subject. The first paper considers fluid mechanics problems associated with interpretation of certain forced oscillation data. Innovative methods for determination of respiratory impedance parameters from forced oscillation data, using nonlinear parameter estimation, forward modeling, and inverse modeling techniques, are presented in the second paper.

The interaction of mass transport with airway mechanics has long been a focus of respiratory and exercise physiology, as well as conventional and high-frequency mechanical ventilation research. Studies during the 1980's underscored the need to understand this interaction at a basic level. The next three papers are concerned with this problem. First, the effect of airway flexibility and geometry (curvature and taper) on mass transport and dispersion are addressed. Next, the interaction of lung mechanics with mass transport mechanisms such as axial streaming, Taylor dispersion, axial diffusion, cardiogenic mixing, and pendelluft are reviewed. Then, recent morphologic studies of the geometry of the alveolar ducts and associated capillary beds are coupled with computer simulation techniques to permit formulation of appropriate transport coefficients for alveolar level mass transport and gas exchange.

Measurements of pressure drop and air flow in the nasal passages have been used routinely for assessing nasal blockage and the efficacy of certain clinical procedures. However, an understanding of the fluid mechanics of this part of the respiratory system is lacking. The last paper presents a fluid mechanics analysis of nasal airflow, including quasi-steady inspiratory and expiratory flows, and methods suitable for describing the flow dynamics at normal and elevated breathing rates.

These papers reflect only a small number of the unsolved problems concerning respiratory gas flows and mass transport. They do, however, delineate several of the more sophisticated approaches in experimental design, computer simulation, and statistical data analysis which are contributing to progress in this area.

IMPEDANCE OF LAMINAR OSCILLATORY FLOW SUPERIMPOSED ON A CONTINUOUS TURBULENT FLOW : APPLICATION TO RESPIRATORY IMPEDANCE MEASUREMENT.

B. Louis and D. Isabey

Unité de Physiologie Respiratoire, INSERM U. 296
Hôpital Henri Mondor, Créteil, France.

1. INTRODUCTION

Measurement of respiratory impedance by Forced Oscillations is an efficient method for characterizing the mechanical behavior of the respiratory system [11, 12]. Recently, this method has been shown to enable early detection of airway abnormalities in subjects exposed to respiratory irritants [2]. The advantage of this non-invasive method is that it does not require cooperation from the patient, allowing him to breathe spontaneously through a connecting tube, usually flushed by a constant bias flow [2,11,15]. This is possible because the oscillating system is opened to the atmosphere by means of a hole or a side tube intended to be a low pass filter for the oscillations [12]. Surprisingly, the interaction between the quasi-steady component (spontaneous breathing and/or bias flow) and the oscillatory component of flow has rarely been considered from a fundamental point of view, except for two recent studies in laminar flow conditions [5,7]. However, a crucial problem remains for the case when the continuous component of flow becomes turbulent : then a linear phenomenon, i.e., the oscillatory flow component [12], is superimposed on a typically non-linear phenomenon, i.e., the turbulent component.

We try in this study to understand such a paradoxical situation from a fluid mechanical point of view, and propose a criterion to predict when the interaction between a quasi-steady turbulent flow component and the superimposed oscillatory laminar flow component might affect the measurement. This study is limited to the interaction between two well-defined types of flow, namely a fully developed, hydraulically smooth, turbulent flow [14] and a laminar oscillatory compressible flow in a long tube [3,6].

2. THEORY

The Stokes-Reynolds number : the problem of the interaction between a steady turbulent component and an oscillatory component has been intensely studied in the engineering field in the last decade. Incidentally, this problem differs from the transition from laminar to turbulent flow during either sinusoïdal or pulsatile flow cycles, which has been more classically studied in physiology [10]. For the simplest cases of the flat plate or the rectangular channel, at least two non-dimensionnal numbers have been considered : the classical Strouhal number $\omega X/U$ (ω : pulsation, X : characteristic length, U : mean velocity of the quasi-steady component) [4] and a new parameter, the Stokes-Reynolds number :

$$ls^{+} = ls/l\nu = (2\,u_{*}^{2}/\omega.\,\nu)^{1/2} \qquad (1)$$

initially proposed by Binder et al [1], which compares the oscillating layer scale or Stokes length $ls = (2.\omega/\nu)^{-1/2}$; ν: kinematic viscosity) to the length scale for viscous effects in turbulent flow, $l\nu = \nu/u_{*}$. u_{*} is the shear velocity, which is classically defined by $u_{*}=(\tau_0/\rho)^{1/2}$ where τ_0 is the wall shear stress. The velocity profile in turbulent flow is known to exibit a logarithmic region and an adjacent linear region close to the wall. The latter is dominated by viscous effects, since turbulent shear stress is negligible near the wall. In an in-depth investigation of the wall region including velocity profile and shear stress measurements in a liquid channel, Binder et al [1] demonstrated that oscillating flow was independent of turbulence if :

$$ls \leq 10.l\nu \quad \text{, i.e.,} \quad ls^{+} \leq 10 \qquad (2)$$

This is an equivalent way of saying that the spanwise vorticity initially produced at the wall by the unsteady component of the pressure gradient is solely removed by the effect of viscous diffusion. This assumption is consistent with the well-known experiments indicating that the wall viscous region extends up to a distance of 5 or $7.l\nu$ [1,4,14].

Application to tube flows : application of this criterion to circular tubes (diameter : D, length : L) is facilitated since the wall shear stress, τ_0 and then the shear velocity, u_{*}, can simply be determined from the longitudinal pressure gradient, $\Delta P/L$, or tube resistance coefficient, Λ :

$$\tau_0 = (\Delta P/L).(D/4) = (\Lambda/4).(1/2.\rho.U^{2}) \qquad (3)$$

U is the mean cross-sectional velocity.

In the case of fully developed, turbulent flow (presently studied), the well-known Blasius resistance formula applies ($\Lambda = 0.32 \cdot Re^{-1/4}$) leading to :

$$u_* = 0.2 \cdot U \cdot Re^{-1/8} \qquad (4)$$

Re = (U.D/ ν) is the Reynolds number of the quasi-steady flow component. Thus, in this precise case, the proposed criteria (equation 2) becomes :

$$Re^{7/8} \leq (100/\surd 2) \cdot \alpha \qquad (5)$$

where $\alpha = (D/2)/(\omega / \nu)^{1/2}$ is the Witzig-Womersley parameter characterizing the superimposed oscillatory flow component.

In order to validate the above criteria (2 or 5) for a gas oscillating in a cylindrical tube, a second non-dimensional parameter , i.e., $\omega . L/c_o$ (c_o : speed of sound) has to be considered in addition to α [3,6]. Accordingly, the entry impedance of a tube opened to the atmosphere, can be expressed :

$$Z_o = (Z/Y)^{1/2} \cdot \text{tgh}[(Z.Y)^{1/2}.L] \qquad (6)$$

where Z and Y are respectively the α-dependent longitudinal impedance and shunt admittance per unit of tube length. Following a method initiated by Van de Woestijne et al [15], the theoretical formula above (6) can be experimentally assessed from the transfer function between two signals of pressure (P_1, P_2), measured over a distance Δx in a measuring device consisting of a circular tube connected upstream to the measured tube, which remains open to the atmosphere.

$$Z_o = \{(Z/Y)^{1/2}.\sinh[(Z.Y)^{1/2}. \Delta x]\}/\{(P_1/P_2) - \cosh[(Z.Y)^{1/2}. \Delta x]\} \qquad (7)$$

This is equivalent to saying that the tested tube acts as a boundary condition for the measuring device.

Entrance effects : although the tested tube was always opened to the atmosphere, it must be emphasized that the boundary conditions of the tested tube (pressure P_e) actually depend on the flow direction as has recently been reported [7]. Thus, when the steady flow component is directed from the tube toward the atmosphere, $P_e = 0$.

However, for a steady flow component directed inward the tested tube, from the atmosphere, $P_e = -\rho U^2$ from momentum theorem. This is equivalent to considering that entrance effects can be represented by adding to the laminar oscillatory flow impedance given by equation (6), a local resistance R such as :

$$R = P_e/\dot{V} = -\ \rho . U/A \qquad (8)$$

where the mean flow $\dot{V}$ (= U.A) is negative since U is negative. This correction for entrance effects is valid only if we neglect the small component of the oscillatory velocity.

3. EXPERIMENTAL METHODS

To validate the above theory, the same 1.6 cm in diameter tube was used as the tested tube and the measuring device. The tested tube was 200 cm long and the measuring device required an additional 17.9 cm (= Δx) of tube length.

The steady flow component in this tube was obtained by sucking air from the atmosphere by means of a vacuum adjustable at various predetermined flow rates, in the range 430 - 1,900 cm^3/s, corresponding to *Re* in the range 2,300 - 10,000. We verified that sucking air from the atmosphere was favorable to rapidly establish a hydraulically smooth turbulent flow. The disadvantage of this method was the inevitable entrance effects, which were nevertheless corrected as described above.

The oscillatory flow component was sinusoidal, created by a loudspeaker connected in parallel to the vacuum. The range of frequency tested was 0 - 250 Hz. The amplitude of oscillatory flow rate at the tube entry varied from 13 to 100 cm^3/s, i.e., representing only from 0.7% to 23% of the steady flow component.

Oscillatory pressure (P_1, P_2) was measured with two ± 140 cmH_20 piezoresistive pressure transducers (Endevco) of 45 kHz resonant frequency, with head located right at the inner wall of the measuring device. Signal were digitized and averaged over 100 cycles in order to decrease the noise to signal ratio. The pressure-flow relationship of the steady flow component was measured with a differential pressure transducer (Validyne) and a screen flowmeter, linear up to 10 l/s.

4. RESULTS AND DISCUSSION

The validity of the laminar oscillatory flow solution (equation 6) was preliminary assessed in the absence of continuous flow (U = 0). Likewise, the agreement between the continuous pressure-flow relationship and the Blasius resistance formula was preliminary verified in the absence of oscillation. The first important result of this study is that the superimposition of oscillations never affected the steady pressure-flow relationship. This result is consistent with those obtained by Cousteix et al [4] or Binder et al [1] who found that neither the mean velocity profile nor the mean turbulent intensity profile were altered by oscillations, even of relatively large amplitude (30% of the mean flow) and whatever values of the Stokes-Reynolds number ls^+ (equation 1). The second important result of this study concerns the comparison between the measured real part of the oscillatory flow impedance at $U \neq 0$, i.e., Real Z, and the basal value obtained at $U = 0$, i.e., Real Z_o, which is presented against ls^+ on the diagram in figure 1. The imaginary part was never modified.

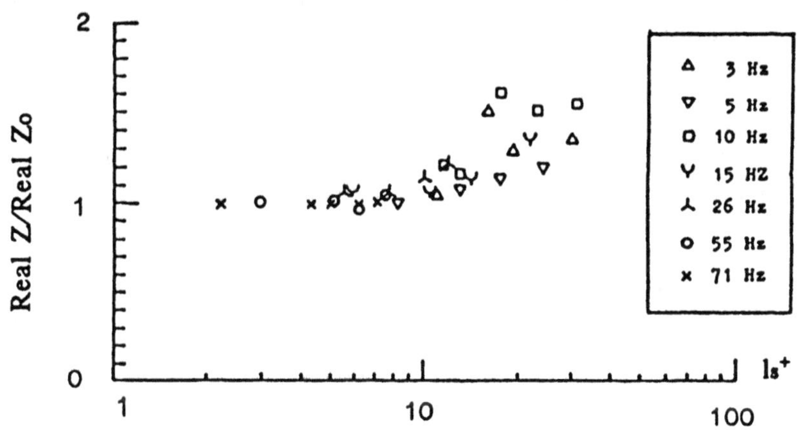

Figure 1. Diagram of the ratio Real Z/ Real Z_o vs ls^+

It is norteworthy to observe that the ratio Real Z/Real Z_o remains close to 1 as long as $ls^+ \leq 10$. This result demonstrates that a small amplitude oscillatory flow superimposed on a non-linear turbulent flow will still behave as a laminar and linear model without modification of the impedance (equation 6) provided that ls^+ is smaller than 10, i.e., as long as the Stokes region remains embedded in the viscous region of turbulent flow. On the contrary, when ls^+ increases above 10, the ratio "Real Z/Real Z_o" rapidly differs from 1, which demonstrates that the tube impedance is drastically modified by the turbulent component. Because the oscillatory flow in the measuring tube is necessarily modified the same way as in the tested tube, it cannot be said that the behavior depicted in figure 1 for $ls^+ > 10$ represents the true change in tested tube impedance, i.e., equation (7) becomes inaccurate. However, the method presently used has the advantage of predicting the occurence of interaction between the steady and the oscillatory tube flow with a high degree of confidence. This latter assertion also requires that entrance effects were correctly taken into account, e.g., the non-linear effects on oscillating and turbulent terms of velocity remain negligible. We assumed that this is true in the conditions of the present study contrary to [7], due to the small relative amplitudes of oscillation presently used. Further, entrance flow during the passage from U = 0 (atmosphere) to turbulent flow was most likely laminar in such systems.

From a fundamental point of view, the criterion proposed by Binder et al [1] was the first to be based on the relative thickness between the Stokes layer (ls) and the viscous sublayer ($l\nu$). Previous criteria were rather based on the comparisons between convective acceleration and local acceleration, i.e., Strouhal number [4], or between frequency of forced oscillation and frequency of the turbulent bursting process [13]. Contrary to ls^+, these latter criteria cannot really explain why dynamic similarity of viscous oscillatory flow phenomena can coexist with dynamic similarity for turbulent flow phenomena. Importantly, the present experiments validate the criteria proposed by Binder et al [1] for gas flow in a tube. Incidentally, the criteria proposed to describe the transition from laminar to turbulent flow in a given sinusoidal cycle [9, 10] are not applicable. For instance, the turbulent counterpart of α [10] leads to assuming unsteady flow for : $Re^{0.5} < 11.5 \cdot \alpha$, which is quite different from criteria (5) and cannot explain the present results.

From a physiological point of view, the present study allows extending the long tube method proposed by Van de Woestijne et al. [15] to forced oscillations superimposed to turbulent steady flow (bias flow or spontaneous breathing flow) provided $ls^+ \leq 10$. This is contradictory to previous recommandations of authors which strictly limit the applicability of such a method to laminar flow conditions ($Re \leq 2300$) and absence of non-linearities [8, 15]. The present results in predicting the interaction between spontaneous breathing and forced oscillations in the airways are applicable to the trachea where flow is known to be turbulent [10]. Below, flow is more laminar and thus the interaction less probable. Taken together these results argue for using high frequencies of oscillations so that $ls^+ \leq 10$ when spontaneous flow rates are high.

REFERENCES

1. Binder, G., Kueny, J.L. : "Measurements of the Periodic Velocity Oscillations Near the Wall in Unsteady Turbulent Channel Flow". Proc. Symp. on Turbulent Shear Flows 3, Univ. of California, Davis (1981).

2. Brochard, L., Pelle, G., De Palmas, J., Brochard, P., Carre, A., Lorino, H., Harf, A. : Density and frequency dependence of resistance in early airway obstruction. Am. Rev. Respir. Dis., **135**, 579-584 (1987).

3. Brown, F.T. : The transient response of fluid lines. J. Basic Engin., **84**, 547-553 (1962).

4. Cousteix, J., Houdeville, R., Javelle, J. : "Response of a Turbulent Boundary Layer to a Pulsation of the External Flow with and without Adverse Pressure Gradient", Proc. IUTAM Symp. on "Unsteady Turbulent Shear Flows", Toulouse, France (1981).

5. Dorkin, H.L., Jackson, A.C., Strieder, D.J., Dawson, S.V. : Interaction of oscillatory and unidirectional flows in straight tubes and an airway cast. J. Appl. Physiol., **52**, 1097-1105 (1982).

6. Franken, H., Clément, J, Cauberghs, M., Van de Woestijne, K.P. : Oscillating flow of a viscous compressible fluid through a rigid tube : a theoretical model. IEEE Trans. Biomed. Engin., **28**, 416-420 (1981).

7. Franken, H., Clément, J., Van de Woestijne, K.P. : Superposition of Constant and Oscillatory flows in a rigid cylindrical tube : influence of entrance effects. IEEE Trans. Biomed. Engin., **33**, 412-419 (1986).

8. Fredberg, J.J., Keefe, D.H., Glass, G.M., Castile, R.G., Frantz III, I.D. : Alveolar pressure nonhomogeneity during small-amplitude high-frequency oscillation. J. Appl. Physiol. **57**, 788-800 (1984).

9. Ohmi, M., Iguchi, M., Kakehashi, K., Masuda, T. : Transition to turbulence and velocity distribution in an oscillating pipe flow. Bull. JSME, **25**, 365-371 (1982).

10. Pedley, T.J., Drazen, J.M. : Aerodynamic Theory. In : *Handbook of Physiology - The Respiratory System III*, P.T. Macklem and J. Meads (Eds), American Physiological Society, Bethesda, pp. 41-54 (1986).

11. Peslin, R. : Méthodes de mesure de l'impédance respiratoire totale par oscillations forcées. Bull. Eur. Physiopathol. Respir., **22**, 621-631 (1986).

12. Peslin, R., Fredberg, J.J. : Oscillation mechanics of the respiratory system. In : *Handbook of Physiology - The Respiratory System III*, P.T. Macklem and J. Mead (Eds), American Physiological Society, Bethesda, pp. 145-177 (1986).

13. Ramaprian, B.R., Tu, S.W. : Fully developed periodic turbulent pipe flow. Part 2. The detailed structure of the flow. J. Fluid. Mech., **137**, 59-81 (1983).

14. Schlichting, H. : *Boundary-Layer Theory*. Translated by J. Kestin, McGraw-Hill Book Company, 817 p. (1979).

15. Van de Woestijne, K.P., Franken, H., Cauberghs, M., Landser, F.J., Clément, J. : A modification of the forced oscillation technique. In : *Adv. Physiol. Sci. Vol. 10. Respiration*, I. Hutas and L.A. Debreczeni (Eds), Pergamon Press, pp. 655-660 (1981).

Current Issues in Understanding Acoustic Impedance of the Respiratory System

Andrew C. Jackson
Respiratory Research Laboratory
Department of Biomedical Engineering
Boston University
Boston, MA 02215

1 Introduction

Most commonly used pulmonary function tests require patients to perform some sort of gymnastic respiratory maneuver such as a forced expiration. These tests are therefore restricted to conscious, cooperative adult humans. Measurements of mechanical, or input acoustic impedance of the respiratory system (*Zin*) can be made in animals, non-cooperative adults and children, and infants. The first measurements of *Zin*, made in normal human subjects were reported by Dubois et al. in 1956 [3]. There were certain characteristics of *Zin* that suggested that the normal respiratory system behaved as though all of the alveoli could be lumped into a single compartment. The single compartment model that they proposed consisted of 6-elements and is illustrated in Fig. 1. If this were the appropriate model and if we could make *Zin* measurements over the appropriate frequency range, there is the potential of estimates of at least two parameters that are of clinical importance, airway resistance (Raw) and thoracic gas volume which is directly related to alveolar gas compression compliance (Cg). At about the same time that DuBois' work was published, Otis et al. [18] reported that *Zin* in patients with diseases of the lung did not behave like a single compartment system. Specifically, they found that the effective compliance of the system was frequency dependent. This implies that there would also be a frequency dependent drop in the real part of *Zin* (Re) for low frequencies; a characteristic not found in normals and one that is not consistent with the 6-element model. Otis et al. suggested that these phenomena were due to inhomogeneities in the parallel airway/tissue pathways. As a result, the respiratory system behaves as though there were two alveolar compartments in parallel, one representing the normal lung tissue, and the other representing the diseased tissue. As a result of these findings, a qualitative feature of the *Zin* spectra, that is a frequency dependence decrease in Re, has been used as an indicator of lung disease. Attempts to extract more quantitative information in the form of specific parameters from *Zin* measurements have only recently been successful.

Since the original publications of DuBois et al. and Otis et al. there have been technological improvements that allow us to make *Zin* measurements to much lower [8] and to much higher frequencies [4], [12]. There have also been considerable improvements in the inverse and forward modelling techniques used to interpret *Zin* data. It is the objective of this work to summarize the results of these inverse and forward modeling studies and to present the current problems associated with our ability to fully understand respiratory system acoustic impedance data.

2 General features of *Zin* spectra.

Examples of *Zin* in a normal dog and human are given in Figs. 2 and 3, respectively. Note that in the dog, there is a frequency dependent drop in Re even in the normal respiratory system but for frequencies below those measured by Dubois et al. ($f < 1$ Hz). A frequency dependent drop in Re at similar frequencies has also been reported in humans [8].

Important features of *Zin* for frequencies above 1 Hz include a resonance (where reactance equals zero) at approx. 10 Hz in dogs and 6 Hz in humans. This resonance is due to the series combination of the summed inertances (Iaw and It) and the tissue compliance (Ct). In dogs there are two antiresonances (where Re has a relative maximum) at approx. 100 and 200 Hz while in humans there is only one at approx. 150 Hz. The first antiresonance in dogs is associated with the parallel combination of the lumped tissue inertance (It) and the gas compression compliance (Cg). The second antiresonance in dogs and the antiresonance in humans is the longitudinal acoustic resonance within the airways. For frequencies below the acoustic antiresonance, the system can be modelled with lumped element models.

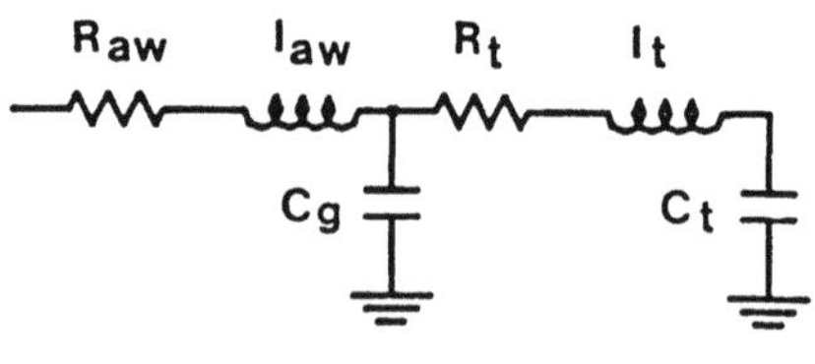

FIGURE 1. Analog circuit representation of 6-element model for the respiratory system suggested by DuBois et al.

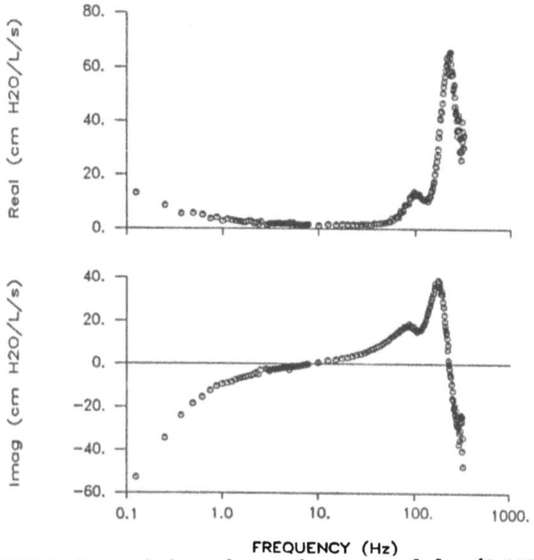

FIGURE 2. Acoustic impedances in a normal dog (0.125 to 320 Hz).

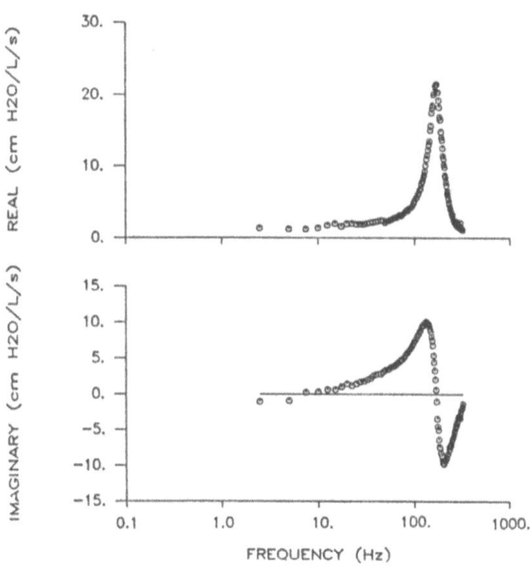

FIGURE 3. Acoustic impedances in a normal human (2.5 to 320 Hz).

a frequency equivalent to that seen in the data. Use this model provides an estimate of the "equivalent length" of the airways. However, this single tube model will not mimic other features of the data. Specifically, the Q-factor of the antiresonance of this model is much larger than that of the data. One mechanism for decreasing the Q-factor of this antiresonance is to represent the airways with multiple parallel tubes all having identical lengths and diameters (Fig. 4). By increasing the number of tubes, the wetted perimeter

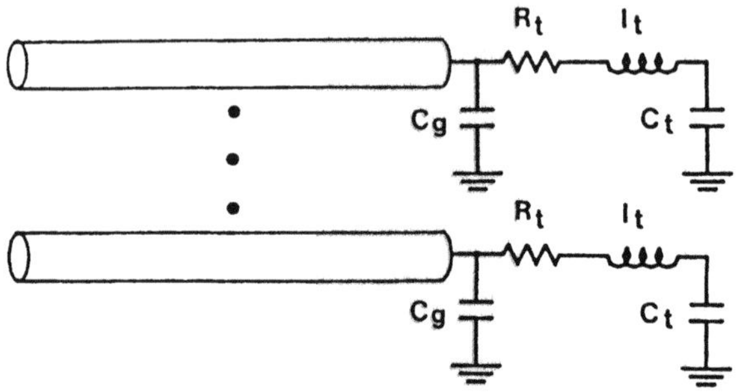

FIGURE 4. Inverse model where airways are represented by multiple, parallel tubes all having the same lengths and diameters. Used byFarre et al [4]. and Lutchen et al [17].

of the system is increased, and the viscous losses are similarly increased. In dogs, these tubes can be terminated with lumped alveolar/tissue elements which provide a mechanism for fitting the "tissue" antiresonance. This multiple tube model fits the dog *Zin* data quite well (Fig. 5) and allows for the extraction of 3 airway parameters, an "equivalent" length, diameter, and number of tubes. This model also fits human data quite well (Fig. 6) but because of the absence of the "tissue" antiresonance, terminal tissue elements are not needed.

One of the problems with using inverse models to interpret *Zin* data has been that even though the respiratory system is an extremely complex system the amount of information (i.e., the number of parameters) that can be extracted from *Zin* spectra is quite limited. As we have seen, the use of linear, lumped element models is limited to frequencies between 1 and about 100 Hz in dogs where the system can be modelled adequately with 6-parameters and to frequencies between 1 and 32 Hz in humans where only 3 parameters can be reliably extracted. For a given frequency range, a more complex model can theoretically be applied resulting in additional parameters. However, there is a limit to the complexity of the model that can be used and this is determined by the complexity of the data (i.e., the number of resonances and antiresonances). A key concern with any inverse

However, to model the data surrounding the antiresonance, distributed parameter models must be used since the wave length of the pressure and flow oscillations within the airways are of the same order of magnitude as the axial dimensions of the airways.

3 Inverse models.

Inverse models have been fit to *Zin* data for frequencies between 0.125 Hz and 320 Hz for the purpose of extracting specific parameters from the data. To model the frequency dependent drop in Re below 1 Hz, two compartment, lumped, linear element models have been used [16]. It was shown that while the data over these frequencies are sufficient to provide statistically reliable parameter estimates, the physiological appropriateness of the estimates is questionable. This is due in part to the fact that in this frequency range the system behaves non-linearly (i.e., impedance is a function of the amplitude of the forcing function). Thus, there is some question as to the usefulness of modeling these data with linear models. At these very low frequencies, *Zin* is dominated by the visco-elastic properties of the lung and chest wall tissues. Models incorporating visco-elastic and plasto-elastic elements have been suggested [6]. But these have not been applied systematically to analyze *Zin* data. A major problem that must be addressed is to separate the frequency dependence of the plasto-elastic properties of normal tissue from the frequency dependence caused by pathological, inhomogeneous parallel pathways.

The respiratory system behaves linearly for frequencies above 1 Hz so these data can be analyzed with linear models. For frequencies surrounding the resonance (i.e, 1 to 32 Hz), *Zin* data can be adequately modelled with only 3 parameters; a single resistance, compliance, and inertance [10]. Consequently, one can hope to reliably extract only 3 parameters from the data (the so-called total respiratory resistance, inertance, and compliance). The 6-element Dubois model represents the next most complex model but before reliable estimates of all can be extracted from the data, impedances that surround the resonance and an antiresonance must be included in the analysis [15]. In dogs, the antiresonance is associated with the lumped alveolar/tissue properties and as a consequence this model is appropriate for use in this species. However, since the antiresonance in humans is not associated with the lumped alveolar/tissue properties this model should not be used to analyze human *Zin* data [17].

The second antiresonance in dogs and the antiresonance in humans is associated with the acoustic properties of the airways [12],[13]. Inverse models to analyze this portion of the *Zin* spectra have been suggested [4],[12], [17]. The simplest of these models is one where the airways are represented by a single rigid walled tube [12]. This model provides a mechanism that has an antiresonance and by adjusting its length it can be made to resonate at

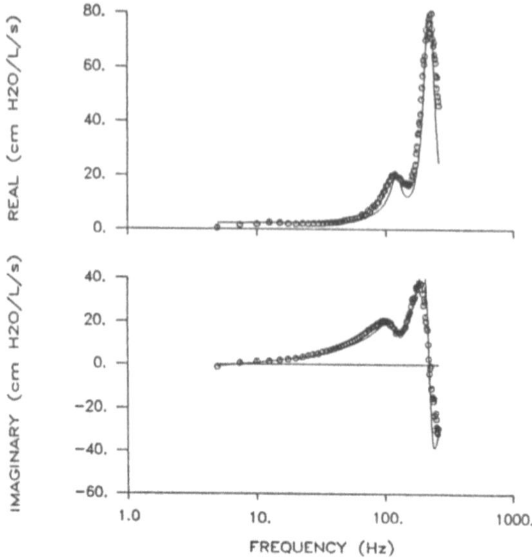

FIGURE 5. Fit of multiple tube model of Fig. [4] to dog acoustic impedances. Equivalent length = 30*cm*, radius = 0.087*cm*, and number of tubes = 117

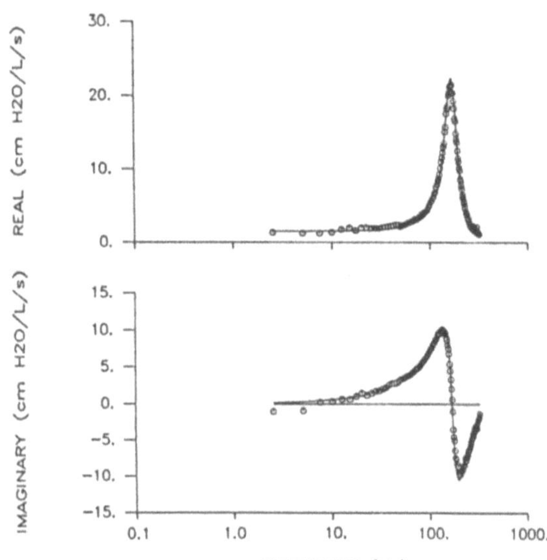

FIGURE 6. Fit of multiple tube model of Fig. [4] (without alveolar/tissue elements) to human acoustic impedances. Equivalent length = 44.7*cm*, radius = 0.075*cm*, and number of tubes = 413

modelling is that as the number of parameters in a model increases, its ability to fit the data may increase but the reliability of the parameter estimates may decrease to unacceptable levels.

4 Forward models.

Forward models that are significantly more complex than these inverse models have been developed in hopes of using them to determine how specific physiological properties (i.e., airway diameters, airway wall properties, and lung/chest wall tissue properties) influence the *Zin* spectra. One of these forward models [9] was based on the Weibel symmetrical human airway model. More recently, These forward models have been based on the recursive asymmetrically branched airway geometries of the human [5] and dog lung [11]. In all of these models, each airway segment is modelled as a transmission line. Equations defining the impedance of circular conduits have been derived by Benade [1]. In applying these equations, however, it must be assumed that the airways are rigid walled and have circular crosssections, that each segment has a constant diameter, and that bifurcations do not contribute significantly to impedance. Ishizaka et al. [9] used their forward model to predict the higher resonance behavior of the human respiratory system. They reported that their model could not mimic *Zin* data surrounding the first acoustic antiresonance unless the airway walls were allowed to be compliant. Similar findings were reported by Fredberg and Hoenig [5] who used an asymmetrically branching model for the airways. Our laboratory has used a forward model of the dog lung to predict the frequency of the first acoustic antiresonance and to then use these data to test the usefulness of the multiple-tube inverse model [11]. Our confidence in the predictions from these forward models is somewhat limited because of the uncertainty in the validity of the underlining assumptions. Probably the two most questionable assumptions are that bifurcations are unimportant and that airway walls are rigid.

4.1 Do bifurcations contribute significantly to *Zin*?

To answer this question experimental and theoretical studies [2] were recently conducted in our laboratory. Measurements of *Zin* were made in a geometrically appropriate cast of the human central airways [7]. This cast consisted of a 4 generation symmetrically bifurcating system of airways with diameters ranging from 2.54 cm in the trachea to 1.14 cm in the most peripheral segments. We found that there were negligible differences between the experimentally measured and theoretically predicted *Zin* for this cast (Fig. 7).

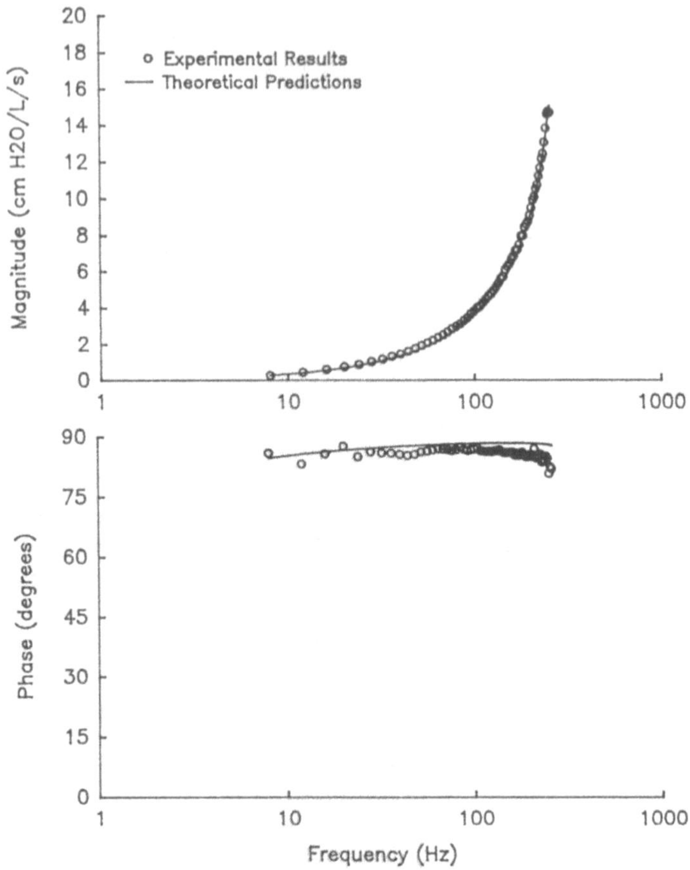

FIGURE 7. Comparison of *Zin* predicted by forward model to measurements of the 4-generation cast of the human central airways.

Evidence from a recent study by Jan et al. [14], suggested that if the ratio of stroke length (stroke length is defined as the average axial displacement of the fluid particle) to the length of the segment is small, the effects of convective acceleration due to the bifurcations can be ignored. This is the case because with small tidal volumes, there is only a small portion of the oscillating fluid particles that travels in the region of the bifurcation and the impedance of the bifurcations per se has a negligible effect. In our study, we predicted the distribution of tidal volumes within the dog airways and found that with the tidal volumes normally used in *Zin* measurements, the stroke length to length ratio is similarly small throughout the entire dog lung. We concluded that our forward models neglecting the impedance of the bifurcation should be expected to provide reasonably accurate impedance predictions.

4.2 Are airway walls non-rigid at these frequencies?

Here we have, perhaps, the most significant issue related to interpreting *Zin* data. Shunting of flow into upper airway wall motion most likely has important influences on *Zin* data at low frequencies (f < 32 Hz) [19]. How airway wall properties influence *Zin* data at higher frequencies is not known but modelling studies suggest that they may be important [5],[9].

5 Summary

The mechanical features of the respiratory system can be identified and separated by formally fitting inverse models to the *Zin* spectra. Analysis of *Zin* data for frequencies surrounding the resonant frequency, distinguishes the gross resistive, inertive, and elastic properties of the total respiratory system. In dogs we can use data surrounding the first antiresonance along with the data surrounding the resonance to separate further the airway properties from the tissue properties. A similar separation of the airway and tissue properties can not be accomplished in humans since the 6-element model can not be used. Analysis of data surrounding the acoustic antiresonance provide a means of extracting additional data relative to the airways. The single-tube model of the airways provides an equivalent length estimate while the multiple-tube model of the airways provides three airway parameters, equivalent length, diameter, and number of tubes. These models represent airway geometry in a rather ad hoc manner and it is not clearly understood how changes in the model parameters can be interpreted in terms of altered airway geometry. Complex forward models have been developed and these may provide insights into how alterations in the numerous properties of the respiratory system effect *Zin*. However, these forward models are far too complex to be used reliably as inverse models. We are currently exploring inverse models that are a compromise between the complex forward models and the simplistic inverse models that have been used. We believe that the best inverse model might be one based on the complex branching models of the airways but one whose complexity is reduced by allowing parameters to vary similarly. For example, allowing the lengths and/or diameters of all airway segments to vary by some constant factor. We currently believe that there is strong evidence that the airway walls must be modelled as compliant structures but we are uncertain about the detailed topology of these models. Even though there remains several important issues, the most crucial of which is probably the issue of the airway wall compliance, inverse modelling of *Zin* data still has the potential of providing quantitative parameter estimates that are of clinical importance.

Acknowledgements: This work was supported by NIH Grant HL-31248.

6 REFERENCES

[1] Benade, A.H. On the propagation of sound waves in a cylindrical conduit. *J. Acoust. Soc. Am.* 44:616-623, 1968.

[2] Bunk, D.A. Modeling forced oscillations in bifurcating airways. M.S. Thesis, Boston University, 1989.

[3] Dubois, A.B., A.W. Brody, D.H. Lewis, and B.F. Burgess, Jr. Oscillation mechanics of lungs and chest in man. *J. Appl. Physiol.* 8:587-594, 1956.

[4] Farre, R., R. Peslin, E. Oostveen, B. Suki, C. Duvuvier, and D. Navajas. Human respiratory impedance from 8 to 256 Hz corrected for upper airway shunt. *J. Appl. Physiol.* 67:1973-1981, 1989.

[5] Fredberg, J.J. and A. Hoenig. Mechanical response of the lungs at high frequencies. *J. Biomech. Eng.* 100:57-66, 1978.

[6] Fredberg, J.J. and D. Stamenovic. On the imperfect elasticity of lung tissue. *J. Appl. Physiol.* 67:2408-2419, 1989.

[7] Hammersley, J.R., B. Synder, and D.E. Olson. Geometric deteminants of airflow distribution within the lung. *Advances in Bioengineering:ASME* 66- 67, 1986.

[8] Hantos, Z., B. Darozy, B. Suki, G. Galgoczy, and T. Csendes. Forced oscillatory impedance of the respiratory system at low frequencies. *J. Appl. Physiol.* 60:128-132, 1986.

[9] Ishizaka, K., M. Matsudaira, and T. Kaneko. Input acoustic impedance measurements of the subglottal system. *J. Acoust. Soc. Am.* 60:190-196, 1976.

[10] Jackson, A.C., J.W. Watson, and M.I. Kotlikoff. Respiratory system, lung, and chest wall impedances in anesthetized dogs. *J. Appl. Physiol.* 57:43-39, 1984.

[11] Jackson, A.C., K.R. Lutchen, and H.L. Dorkin. Inverse modeling of dog airway and respiratory system impedances. *J. Appl. Physiol.* 62:2273-2282, 1987.

[12] Jackson, A.C., C.A. Giurdanella, and H.L. Dorkin. Density dependence of respiratory system impedances between 5 and 320 Hz in humans. *J. Appl. Physiol.* 67:2323-2330, 1989.

[13] Jackson, A.C. and K.R. Lutchen. Physiological basis for resonance frequencies of respiratory system impedances in dogs. *Submitted for publication.*

[14] Jan, D.L, A.H. Shapiro, and R.D. Kamm. Some features of oscillatory flow in a model bifurcation. *J. Appl. Physiol.* 67:1470159, 1989.

[15] Lutchen, K.R. and A.C. Jackson. Reliability of parameter estimates from models applied to respiratory data. *J. Appl. Physiol.* 62:403-413, 1987.

[16] Lutchen, K.R., Z. Hantos, and A.C. Jackson. Importance of low-frequency impedance data for reliably quantifying parallel inhomogeneities of respiratory mechanics. *IEEE Trans. Biomed. Eng.* 35:472-481, 1988.

[17] Lutchen, K.R., C.A. Giurdanella, and A.C. Jackson. Inability to separate airway from tissue properties using human respiratory input impedance. *J. Appl. Physiol.*. In press. 1990.

[18] Otis, A.B., C.B McKerrow, R.A. Bartlet, J. Mead, B.B. McIlroy, N.D. Selverstone, and E.P. Radford, Jr. Mechanical factors in the distribution of pulmonary ventilation. *J. Appl. Physiol.* 8:587-594, 1956

[19] Peslin, R., C. Duvivier, C. Gallina, and P. Cervantes. Upper airway artifact in respiratory impedance measurements. *Am. Rev. Resp. Dis.* 132:712-714, 1985.

Effects of Curvature, Taper and Flexibility on Dispersion in Oscillatory Pipe Flow

James B. Grotberg, Ph.D., M.D.
Biomedical Engineering Department
Northwestern University
Evanston, IL 60208

Introduction

The primary function of the lung is the exchange of oxygen and carbon dioxide. Other substances, such as anesthetics, aerosols, and toxins, are delivered and removed from the lungs by a similar process. Mass transport in the lung depends on convection, diffusion, and their interaction during oscillatory flow. High frequency ventilation (HFV) is an alternative type of ventilation under investigation (Bohn et al. 1980). In HFV, relatively small tidal volumes of air (35-150 ml) are delivered at high frequencies (5-30 Hz), so that pulmonary barotrauma and cardiac impairment are avoided.

A theoretical model of the lung undergoing HFV would lend insight to the important mechanisms and geometrical features that enhance or inhibit mass transport, leading to the improved design of ventilators and to the development of diagnostic techniques. The simplest model of a pulmonary airway is a straight, rigid tube. Chatwin (1975) and Watson (1983) studied mass transport during oscillatory flow in such a model, and Watson's theoretical results were verified experimentally by Joshi et al. (1983). The transport rate varies with the Womersley parameter, $\alpha=a(\omega/\nu)^{1/2}$; the amplitude parameter $A=d/a$; and the Schmidt number $Sc=\nu/D$. Here, a is the tube radius, ω the angular frequency of oscillation, and ν the kinematic viscosity of the fluid. The stroke distance is defined by $d=V_T/\pi a^2$, where V_T is the tidal volume, and D is the molecular diffusivity of the transported substance.

Experimental results of HFV (Bohn et al. 1980, Rieke et al. 1983, and Mitzner et al. 1983), suggest the existence of a frequency of oscillation which optimizes gas exchange, a result not predicted by the straight, rigid tube theories. Indeed, the lung's airways consist of tapered, curved, branching, flexible-walled tubes of varying length and diameter.

Each geometrical and physical feature may enhance or inhibit mass transport at a particular tidal volume or frequency of oscillation. A reasonable first approach is to examine each feature individually as a means of understanding the total phenomenon.

Problem Formulation

We have investigated the effects of taper, axial curvature and flexibility individually. In general, the flow field must be determined from the Navier-Stokes equations for the appropriate geometry and boundary conditions.

$$\frac{\partial \mathbf{U}}{\partial t} + (\mathbf{U}\cdot\nabla)\mathbf{U} = -\frac{1}{\rho}\nabla P + \nu\nabla^2\mathbf{U} \qquad (1)$$

where the kinematic boundary conditions on the velocity vector, **U**, are imposed as no slip at the tube wall and finite velocity at the tube centerline. The imposed pressure gradient is taken to be time-periodic of sufficient amplitude to keep the tidal volume fixed. Once the flow field has been determined, it must be inserted in the convection-diffusion equation which describes the transport of material,

$$\frac{\partial C}{\partial t} + (\mathbf{U}\cdot\nabla)C = D\nabla^2 C \qquad (2)$$

subject to zero mass flux at the tube wall and finite concentrations at the tube midline. In the case of a flexible wall there is the additional stress boundary condition which relates the fluid stresses to the wall stresses and wall acceleration.

We fix the time-averaged concentration at two locations along the tube axis, say at z=0 and z=L. Then, in the absence of convection, a steady concentration gradient, $\Delta C/L$, exists. Imposing the volume-cycled oscillatory flow results in transport enhanced by the interaction of convection and diffusion. A measure of that transport is the mass flow rate, $\overset{*}{m}$, averaged over a complete cycle,

$$\overset{*}{m} = \frac{\omega}{2\pi}\int_{t_0}^{t_0+\frac{2\pi}{\omega}}\int_0^{2\pi}\int_0^{R(z,t)}[WC - D\frac{\partial C}{\partial z}]\ r\ dr\ d\theta\ dt \qquad (3)$$

where W is the axial component of velocity. R(z,t) is a constant for the curved tube, a function of z only for the tapered tube, and a function of z and t (determined in the analysis) for the flexible tube.

Results

The effects of taper on mass transport were studied theoretically by Grotberg (1984) and Godleski and Grotberg (1988) and experimentally by Gaver and Grotberg (1986). Assuming a small taper angle, ϵ, Godleski and Grotberg found solutions for the velocity and concentration fields that were regular perturbation expansions in ϵ, essentially a lubrication theory. See Figure 1 where the dimensionless mass flow rate, $m=\overset{*}{m}/(\nu a\Delta C)$, is plotted versus α for A=5.0, Sc=0.9 and $\epsilon=0$ (straight tube), 0.03 and 0.1 as indicated. The radial scale, a, is taken at the small end and L/a=13. For $\alpha<3$, the better transport is given by the more strongly tapered tube, while for $\alpha>3$ the straight tube yields better transport.

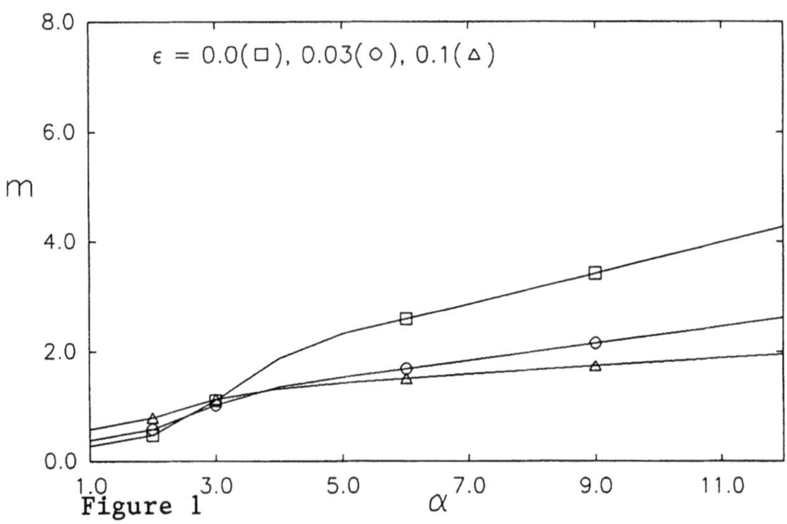

Figure 1

Eckmann and Grotberg (1988) investigated the effect of tube curvature on axial mass transport. In their analysis, $\delta=a/R_0\ll1$, where R_0 is the radius of axial curvature. Solutions to the Navier-Stokes and convection-diffusion equations in toroidal coordinates were posed as regular perturbation expansions in δ. See Figure 2 for $\delta=0$ (straight tube) and $\delta=0.3$. Again we let A=5.0, Sc=0.9 and L/a=13. This figure shows that tube curvature enhances the rate of axial mass transport for all values of A and α. For this value of A, a local maximum value of mass transport occurs near $\alpha=5$, followed by a local minimum. Indeed, there is little change in m for $5<\alpha<9$, approximately. This local extremum occurs because of the phase relationship between the velocity and concentration fields which is affected by the swirling motions induced from the curvature.

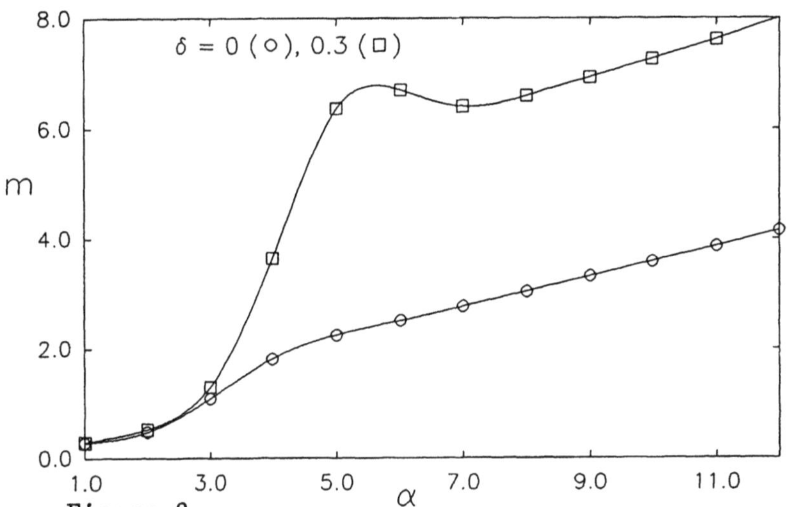

Figure 2

The effect of wall flexibility on gas exchange efficiency is particularly important, since, unlike curvature or tapering, airway flexibility changes in diseased states. In the lungs, wall flexibility can be interpreted as compliance, which is defined as the airway volume change per unit pressure change. Patients with fibrosis and infant respiratory distress syndrome suffer from decreased compliance. Aging and emphysema tend to increase lung compliance. Understanding the relationship between compliance and gas transport during HFV is essential; a given stroke amplitude and frequency may adequately ventilate one patient, while a completely different set of parameters would be necessary for another.

Establishing the dependence of gas exchange efficiency on lung compliance could also enable HFV to be used as a diagnostic tool.

Considering a long-wave, small amplitude limit, Dragon and Grotberg (1989), analyzed oscillatory flow in a flexible tube employing a visco-elastic tube model after Atabek and Lew (1966). The oscillatory pressure gradient causes fluid-elastic waves to propagate as the tidal volume reverses. Figure 3 shows the mass flow rate, m, as a function of α for A=5.0 and Sc=0.9, as in the two previous examples. Three new parameters emerge in this model: the damping ratio $G=ga^2/(2\nu)$ where g is the wall damping; the mass ratio, $M=\rho_w h/(\rho a)$, where ρ_w is the wall density and h is the wall thickness; and, the wave speed ratio, $\kappa=Ea^2/(\rho_w \nu^2)$ where E is the Youngs modulus. $\kappa^{1/2}$ is a ratio of a characteristic elastic wave speed, $(E/\rho_w)^{1/2}$, to a characteristic speed of shear propagation through the viscous fluid, ν/a. It is κ that reflects the compliance of the airways which is inversely relate to E. In Figure 3 we have m plotted against α for A=5.0, Sc=0.9, M=49, G=13.4, L/a=13 and four values of κ as shown.

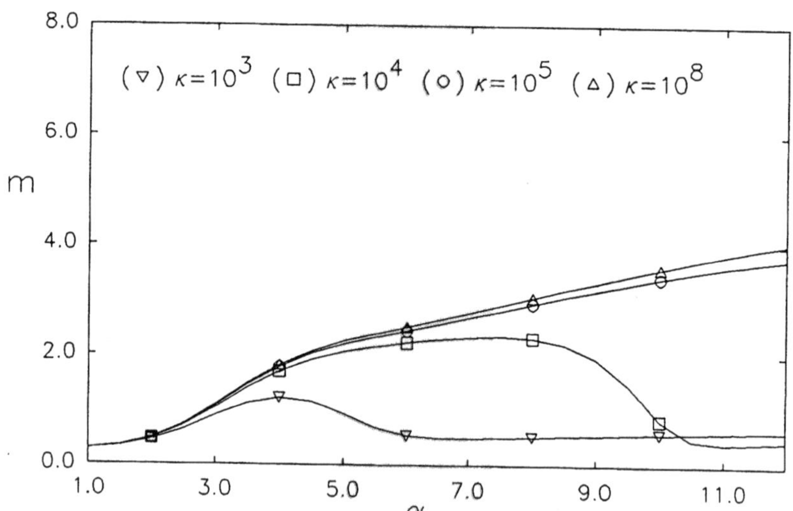

Figure 3

We first note that $\kappa = 10^8$ is essentially the rigid tube case over this range of α and gives the same result for rigid straight tubes seen in the previous figures. All of the values of m fall below the rigid tube limit, so clearly wall flexibility reduces transport. Most striking, however, is the appearance of a maximum mass flow rate for $\kappa=10^3, 10^4$. The amplitude of the maximum decreases with κ and occurs at ever smaller values of α. When the local wall motion is analyzed we find that the phase relationship between the wall position and the volume flow rate achieves a minimum at approximately the same values of α where the local maximum in m occurs.

The presence of the local maximum in the m versus α curves may explain, in part, the experimental results of Rieke (1983), Mitzner (1983), and Rossing (1981), in which gas exchange efficiency peaked or plateaued with increasing frequency of oscillation. In a clinical setting it is important to know whether to attribute such behavior to pathology or to the complex mass transport characteristics of a normal lung.

In an analysis of the non-linear wave mechanics, Dragon and Grotberg show that steady, bi-directional streaming flows are induced at higher order and there is an induced steady pressure gradient associated with these flows. The steady pressure increases in the direction of the wave, toward the distal end of the tube. A steady pressure gradient (and bi-directional streaming) was also predicted by Grotberg (1984) as a result of taper, and again the steady pressure was found to increase toward the distal (wider) end. These two mechanisms are consistent with the steady pressure gradient observed by Simon, Weinmann and Mitzner (1984) which cause the hyperinflation of a lung during HFV.

References

Atabek, H.B. & Lew, H.S. Biophys. J. 6: 481-503, 1966.

Bohn, D.J., Miyasaka, K., Marchak, E.B., Thompson, W.K., Froese, A.B. Bryan, A.C. J. Appl. Physiol. 48: 710-716, 1980.

Chatwin, P.C. J. Fluid Mech. 71: 513-527, 1975.

Dragon, C.D. and J.B. Grotberg. American Physical Society, Division of Fluid Dynamics Annual Meeting, Palo Alto, California, November, 1989.

Eckmann, D.M. & Grotberg, J.B. J. Fluid Mech. 188: 509-527, 1988.

Gaver III, D.P. & Grotberg, J.B. J. Fluid Mech. 172: 47-61, 1986.

Godleski, D.A. & Grotberg, J.B. J. Biomech. Engng. 110: 283-291, 1988.

Grotberg, J.B. J. Fluid Mech. 141: 249-264, 1984.

Joshi, C.H., Kamm, R.D., Drazen, J.M. & Slutsky, A.S. J. Fluid Mech. 133: 245-254, 1983.

Mitzner, W., Permutt, S. & Wienmann, G. Ann. Biomed. Eng. 11:61, 1983.

Rieke, H., Hook, C. & Meyer, M. Respir. Physiol. 54: 1-17, 1983.

Rossing, T.H., Slutsky, A.S., Lehr, J.L., Drinker, P.A., Kamm, R. Drazen, J.M. N. Engl. J. Med. 305: 1375-1379, 1981.

Simon, B.A., Weinmann, G.G., & Mitzner, W. J. Appl. Physiol. 57: 1069-1078, 1984.

Watson, E.J. J. Fluid Mech. 133: 233-244, 1983.

INTERACTIONS BETWEEN LUNG MECHANICS AND GAS TRANSPORT. James S. Ultman, Department of Chemical Engineering, Penn State University, University Park, PA, 16802, USA.

Lung mechanics, in the broadest sense, encompasses both the behavior of solid tissue and the dynamics of gas and liquid flow within the tissue. Gas transport, which refers to the simultaneous bulk flow and diffusion of gases between the airway opening and the alveolar membranes, interacts with lung mechanics in two ways: the movement of gas molecules within each airway branch and airspace is affected by local velocity fields; and the blending of molecules at airway branch points depends on the distribution of flow imposed by the tidal deformation of elastic tissue. Nearly 100 years ago, in a classic paper that set the stage for comtemporary studies of these processes, Bohr (1) postulated that transport of respiratory gases could be modeled in a lung consisting of a proximal *dead space* compartment terminating in a blind-ended *alveolar* compartment. Gas was assumed to translate through the rigid dead space by bulk flow in the absence of gas mixing, while gas in the distensible alveolar compartment was perfectly mixed. This simple two-compartment model has become an incredibly important tool in the conceptualization of lung function, and is the underlying basis of routine clinical tests such as the single-breath nitrogen washout and the carbon monoxide diffusing capacity tests.

Any mathematical model represents an idealization from which departures in behavior can be recognized, leading eventually to improved models which extend our basic understanding. An important example of this for the Bohr model is its prediction that tidal volume must exceed dead space volume if alveolar ventilation is to occur. In 1915, Henderson et al. (2) used smoke to visualize gas movement in a glass tube. They observed that centrally-located smoke particles were displaced further downstream than would be predicted from the total volume of air introduced into the tube, and hypothesized that the same axial mixing of gas in airway branches could lead to alveolar ventilation even at tidal volumes below dead space volume. With the availability of high speed gas analyzers to accurately measure alveolar gas composition in expired breaths, Briscoe et al. (3) were able to confirm this hypothesis. Employing "helox" as a breathing mixture, they found that alveolar ventilation varied from 10 to 47 ml as inspired tidal volume was changed from 60 to 150 ml in a human subject whose dead space volume was 170 ml.

These results suggest that at a significantly reduced tidal volume, lung parenchyma can still be ventilated with normal minute volumes of fresh air, provided that there is a compensatory increase in breathing frequency. This is the basis of high frequency mechanical ventilation, which offers the clinical advantage of lower driving pressures relative to conventional mechanical ventilation. Much of research into high frequency ventilation explored the mechanisms by which small tidal volume ventilation occurs, and served as a vehicle to improve our understanding of gas transport processes under a broad range of breathing conditions (4). In the remainder of this article, a number of these mechanisms are briefly discussed.

Direct alveolar ventilation is a consequence of the asymmetric branching pattern of the tracheobronchial tree. This results in a non-uniform distribution of transit times between the airway opening and the respiratory airspaces, such that more proximal alveoli receive a greater proportion of fresh air than more distal alveoli. Horsfield and Cumming (5) demonstrated that this effect makes a large contribution to the apparent mixing between dead space and alveolar gas observed during a single breath nitrogen washout test.

Axial streaming, *Taylor dispersion*, and *axial diffusion* are mechanisms of mixing generated by the interaction of diffusion with the non-uniform axial convection of gas molecules within individual airway branches. Gomez (6) pointed out that, because the summed cross-sectional area of airway generations dramatically increases with distance from the airway opening, the comparative importance of convection and diffusion systematically decreases with increasing penetration of fresh air into the tracheobronchial tree. Wilson and Lin (7) built on this concept as well as on G.I. Taylor's earlier studies of mixing in straight tubes (8) and Weibel's 23 generation symmetric branching model (9) to subdivide the lung into three mixing regions. Proximal to the eighth generation, they predicted that mixing occurs by axial streaming, a convectively-driven nonuniform translation of gas in the absence of diffusion. Between the eighth and twelfth, they speculated that mixing occurred primarily by Taylor dispersion, which represents a local balance between axial convection and radial diffusion. And distal to the twelfth generation, they theorized that mixing occurs by axial diffusion alone.

Noninvasive measurements of the longitudinal mixing of insoluble indicator gases have been made by the bolus-response technique in the conducting airways of human subjects (10). In both upper and central airways, mixing was inversely related to the diffusion coefficient of the indicator gas. This is consistent with the action of Taylor dispersion, but not of axial streaming and axial diffusion. On the other hand, mixing in the central airways was independent of respiratory flow suggesting that axial streaming is more important than Taylor dispersion or axial diffusion. In the upper airways, mixing was directly related to expiratory flow and independent of inspiratory flow, implying that there is a shift in mixing mechanisms between inhalation and exhalation. The departure of these data from Wilson and Lin's simple catagorization of mixing mechanisms is due to complexities in airway gas dynamics that do not arise in straight tubes.

An interesting illustration of such a complexity is flow through the larynx, where the the acceleration of gas through the vocal cords generates a confined jet. In the high shear region at the boundary of this jet, turbulence can be generated at tracheal air flows one-tenth as large as those which are necessary to produce fully-developed turbulent flow in an unobstructed tube (11). Simone and Ultman (12) measured the distribution of axial mixing during steady flow through a rigid larynx cast mounted in a straight tube. Immediately downstream of the larynx there was a sudden increase in mixing rate due to the steep velocity profile in the confined jet. The magnitude of this increase was inversely related to the diffusion coefficient of the indicator gas which is suggestive of a Taylor-type dispersion process. Further downstream of the constriction, the mixing rate leveled off at a value below that measured in an unconstricted tube. Undoubtedly this is due to the generation of turbulence by the jet (Paradoxically, the blunting of the velocity profile and the intensification of radial transport by eddy diffusion leads to less axial mixing during turbulent flow than is present during laminar flow.). Where airway constriction is produced by pathological conditions such as bronchial inflammation, mucus hypersecretion, or bronchospasm, it is conceivable that local mixing patterns would be disturbed as they are by the vocal cords (13).

Flow pulsations with amplitudes as large as 2.5 liters per minute and synchronous with the heart beat have been measured in lobar and sublobar bronchi (14), and, theoretically, this should result in significant *cardiogenic mixing* (15). Fukuchi et al (16) were the first to quantify longitudinal mixing due to mechanical interactions of the heart with the lungs. By measuring the flow necessary to produce a "stationary nitrogen front" between inspired oxygen and residual air, these investigators demonstrated that mixing in 2.5-8.6 mm airways of open-chested dogs is more than five times greater during life than post-mortem. However, it is not clear from these data which region of the lung is most affected by the heart.

Using the noninvasive bolus-response technique, Drechsler and Ultman were unable to detect an increase in longitudinal mixing in any region of the airways when heart rate was increased by exercising the subjects (17). This is consistent with the findings of Horsfield et al.(18), who used a cardiac pacemaker to regulate the heart rate of anesthetized dogs, and found no effect on mixing in the living animal. However, by employing a simulation-based analysis of bolus-response data obtained during breathholding, Larsen et al. (19) were able to characterize the distribution of cardiogenic mixing from the mouth to a penetration volume of 240 ml in the lungs of human subjects. Expressed on the basis of a mixing coefficient, which has a value of 0.25 cm^2/sec for the diffusion of oxygen in air, cardiogenic mixing was negligible at the shallowest and deepest penetration volumes, but it exhibited a maximum value of 10 cm^2/sec at an intermediate penetration volume of 130 ml. This penetration roughly corresponds to the fourteenth airway generation.

Pendelluft, the tidal flow of gas between lung units with different mechanical properties was first offered as an explanation for "uneven alveolar ventilation" (20). The Otis mechanical model, the first mathematical analysis of pendelluft, consists of a parallel arrangement of two resistance-compliance paths, $(RC)_1$ and $(RC)_2$. Presumably, this model is equivalent to two Bohr models in parallel where R_1 and R_2 are the resistances to gas flow in the dead space compartments and C_1 and C_2 are the tissue compliances in the alveolar compartments. When $(RC)_1$ and $(RC)_2$ have different values, both the amplitude and the phase of tidal flows delivered to the two alveolar compartments are different.

Consider, for example, a single-breath nitrogen test carried out in a lung for which $R_1 < R_2$ while $C_1 = C_2$. During inspiration, because of its lower resistance to gas flow, path 1 will be inflated with more pure oxygen than path 2. Therefore, residual nitrogen is at a lower concentration in path 1 than in path 2 at the end of inspiration. During expiration, the path with the shorter RC constant, path 1, will empty earlier. This implies that the alveolar plateau will be initially composed of low concentration gas from path 1, but will gradually rise as more concentrated gas from path 2 is expired later during expiration.

An important property of the Otis parallel path R-C model is that the total alveolar ventilation (i.e., the sum of the tidal volumes delivered to the two compliance compartments) is greater than the tidal volume within the trachea. This enhancement in alveolar ventilation, which has a maximum value of 41.4%, can help maintain normal gas exhange in conditions such as hyaline membrane disease or diaphagmatic hernia where there are severe interregional aberrations in mechanics. This property might also help compensate for the small tidal volumes delivered during high frequency ventilation.

Because of the gas inertia generated at high frequency breathing frequencies, the Otis model must be modified for this situation by the addition of inertance elements, I, along each of the two airway paths. In such a R-I-C model, the enhancement in alveolar ventilation is not limited to 41%. In fact, in a tube bifucation with mechanical properties similar to those in a dog lung, enhancements in alveolar ventilation as large as 175% have been observed at an oscillation frequency of 20 Hz (21). One of the interesting features of the parallel path R-I-C model is that for a given set of values for the six mechanical constants, there is always an optimal frequency at which ventilation enhancement is optimized. This optimum ventilation frequency also corresponds to an optimal enhancement of longitudinal mixing observed in hardware models as well as to an optimal enhancement of gas exchange observed in animal experiments (22).

In conclusion, the interaction of fluid mechanics and tissue mechanics with the transport of respired gases occurs in a variety of ways depending on breathing conditions, lung geometry and gas properties. We may be able to capitalize on some of these processes in the design of more sensitive and specific tests of pulmonary function. Also, a fuller understanding of the interaction of mechanics and transport may lead to an improved interpretation of existing function tests.

References

1. Bohr C. 1891. Uber die Lungenatmung. Skand Arch Physiol.

2:236-268.
2. Henderson Y, Chillingworth FP, Whitney JL. 1915. The respiratory dead space. Am J Physiol. 38:1-19.
3. Briscoe WA, Forster RE, Comroe JH. 1954. Alveolar ventilation at very low tidal volumes. J Appl Physiol. 7:27-30.
4. Chang HK. 1984. Mechanisms of gas transport during ventilation by high-frequency oscillation. J Appl Physiol. 56:553-563.
5. Horsfield K, Cumming G. 1968. Functional consequences of airway morphology. J Appl Physiol. 24:384-390.
6. Gomez DM. 1965. A physico-mathematical study of lung function in normal subjects and in patients with obstructive diseases. Med Thorcac. 22:275-294.
7. Wilson TA, Lin K. 1970. Convection and diffusion in the airways and the design of the bronchial tree. In: Airway Dynamics, Bohuys A, ed, Springfield, IL, Thomas, pp 5-19.
8. Taylor GI. 1954. Dispersion of soluble matter in solvent flowing slowly through a tube. Proc Roy Soc Lond. 219A:186-203.
9. Weibel ER. 1963. Morphology of the Lung, New York, Academic Press, ch 11.
10. Drechlser Parks DM, Larsen RW, Ultman JS. 1985. Inert gas mixing in the upper and central airways of man. Respir Physiol. 62:305-324.
11. Dekker E. 1961. Transition between laminar and turbulent flow in human trachea. J Appl Physiol. 16:1060-1064.
12. Simone AF, Ultman JS. 1982. Longitudinal mixing by the human larynx. Respir Physiol. 49:187-203.
13. Weaver DW, Ultman JS. 1980. Axial dispersion through tube constrictions. Am Inst Chem Engr J. 26:9-15.
14. West JB, Hugh-Jones P. 1961. Pulsatile flow in the bronchi caused by the heart beat. J Appl Physiol. 16:697-702.
15. Slutsky AS. 1981. Gas mixing by cardiac oscillations: a theoretical quantitative analysis. J Appl Physiol. 51:1287-1293.
16. Fukuchi Y, Roussos CS, Macklem PT, Engel LA. 1976. Convection, diffusion and cardiogenic mixing of inspired gas in the lung, and experimental approach. Respir Physiol. 26:77-90.
17. Drechsler DM, Ultman JS. 1984. Cardiogenic mixing in the pulmonary airways of man? Respirat Physiol. 56:37-44.

18. Horsfield K, Gabe I, Mills C, Buckman M, Cumming G. 1982. Effect of heart rate and stroke volume on gas mixing in dog lung. J Appl Physiol. 53:1603-1607.

19. Larsen RW. Ultman JS. 1986. Longitudinal mixing coefficients in human lungs. Proc Ann Conf Engr Med Biol. 28:154(Abst).

20. Otis AB, McKerrow CB, Bartlett RA, Mead J, McIlroy MB, Selverstone NJ, Radford EP. 1956. Mechanical factors in distribution of pulmonary ventilation. J Appl Physiol. 8:427-443.

21. Cooke KA, Karl SR, Ultman JS. 1987. Pendelluft flow during high frequency oscillation in a single bifurcation. Proc Ann Conf IEEE Engr Med Biol Soc. 4:2044-2055.

22. Ultman JS, Shaw RG, Fabiano DC, Cooke KA. 1988. Pendelluft and longitudinal mixing in a single-bifurcation lung airway model during high frequency oscillating flow. J Appl Physiol. 65:146-155.

Nonclassical Features of Gas Transport and Exchange at the Alveolar Level

William J. Federspiel
Department of Biomedical Engineering
Boston University
Boston, MA 02215

1 Introduction

The transport and exchange of oxygen in the lung can be viewed conceptually as serial processes. Accordingly, the conductance for the combined transport and exchange process, G_c, represents the reciprocal sum of separate conductances for intrapulmonary mixing, G_{mix}, and for alveolar capillary exchange, D_L [9]. Although G_{mix} and D_L are lumped-paramèter macroscale conductances, they must account implicitly for transport and exchange complexities at the microscale (i.e. alveolar duct and alveolar capillary) level.

This paper describes mathematical and computational studies aimed at examining how microstructural geometric features of the pulmonary transport and exchange apparatus might direct the macroscale conductances for intrapulmonary mixing and alveolar capillary exchange. Primary interest was in incorporating previously neglected geometric features of acinar ducts and capillary blood flow which might impact on overall respiratory transport.

2 Intrapulmonary Gas Mixing

Intrapulmonary mixing between inspired gas and resident alveolar gas occurs predominantly across a relatively stationary inspiratory front where convective and mixing processes come into balance (cf. [8]). Accordingly, G_{mix} should vary in proportion to the axial mixing coefficient, D^*, along the stationary front, and to the cumulative cross-sectional area there. Even with parallel inhomogeneities, gas mixing along a pathway should be proportional to the axial mixing coefficient at the stationary front established for that pathway, and overall G_{mix} will be a complicated average among the individual pathway processes.

The stationary front normally occurs within the pulmonary acinus [8]. In simple straight tubes, the axial mixing coefficient, D^*, is bounded below by the molecular diffusion coefficient, D, and above by the Taylor flow enhanced mixing coefficient [11]. Furthermore, for Peclet numbers (Pe) relevant to the pulmonary acinus (Pe < 1), the straight tube model predicts that axial mixing in acinar ducts would be predominantly by axial diffusion, and $D^* \cong D$. The microscale structure of acinar ducts, however, departs appreciably from simple straight tubes because of alveolarization. Since the axial mixing coefficient is a reflection of microscale phenomena (convection and diffusion at a local level), the impact of acinar duct morphology on the axial mixing coefficient, D^*, must be assessed.

We looked at axial mixing theoretically [6] in a spatially periodic model of an alveolated duct. The axisymmetric model consisted of a cylindrical central conduit embraced by alveoli with circular shapes in longitudinal cross-section (Fig. 1). The local transport of a gas within the duct is gov-

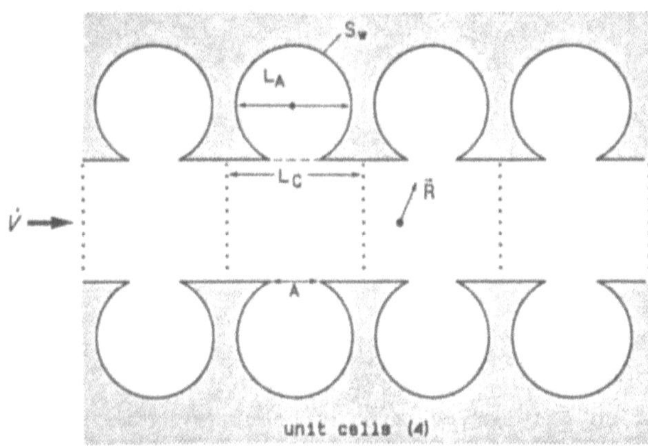

FIGURE 1. Geometric model of alveolated duct.

erned unambiguously by a three-dimensional convection-diffusion equation for gas concentration, $C(\vec{R}, t)$:

$$\partial C/\partial t + \vec{V}(\vec{R}) \cdot \nabla C = D\nabla^2 C \ , \tag{1}$$

where, $\vec{R}$ is position within the duct, t is time, and $\vec{V}(\vec{R})$ is the velocity vector field within the alveolated duct. Eq. 1 represents a microscale description of transport within the alveolated duct. In principle its solution provides gas concentration at all points within the complex domain and gives a complete description of transport.

The axial dispersion coefficient, D^*, is a macroscopic coefficient describing how local convection and diffusion embodied in Eq. 1 govern the axial mixing of average gas concentration along the duct. A moment analysis technique developed by Brenner [2] for spatially periodic porous media was used to transform the microscopic description without loss of rigor to the macroscale axial mixing coefficient, D^* [6]. Using the Brenner methodology, D^* was computed from an integral over a single unit cell (uc) of the periodic structure:

$$D^* = D \int_{\mathrm{uc}} \nabla B \cdot \nabla B \, dV \tag{2}$$

where B is a scalar field defined by a convection diffusion type equation which arises from the moment analysis (the B equation resembles the steady state form Eq. 1 with a uniform source term of magnitude $\bar{V}$, and special periodic boundary conditions). All governing unit cell equations were solved numerically using standard finite difference techniques applied to a numerically generated boundary-fitted coordinate system (grid) [12] which conformed to the complex shape of a unit cell (Fig. 2, left panel).

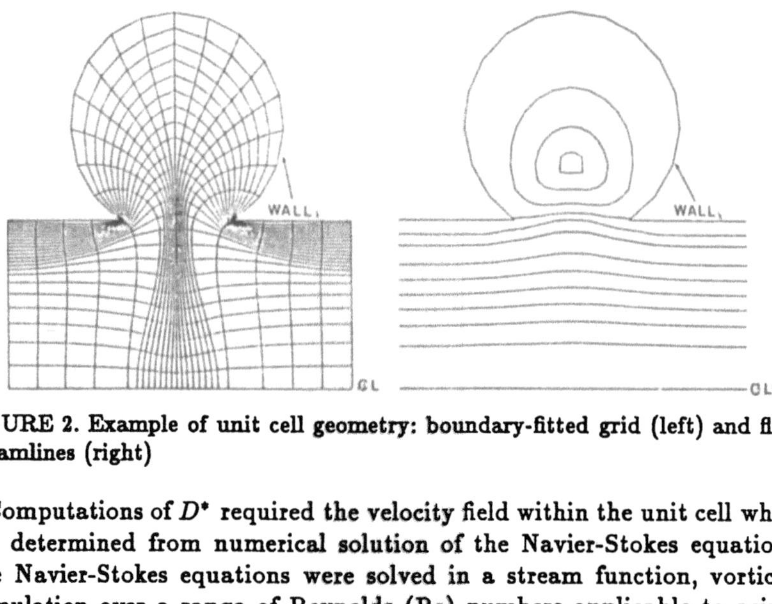

FIGURE 2. Example of unit cell geometry: boundary-fitted grid (left) and fluid streamlines (right)

Computations of D^* required the velocity field within the unit cell which was determined from numerical solution of the Navier-Stokes equations. The Navier-Stokes equations were solved in a stream function, vorticity formulation over a range of Reynolds (Re) numbers applicable to acinar ducts. An example of fluid streamlines for one geometry is shown in Fig. 2 (right panel). Flow patterns were relatively insensitive to Re (results in Fig. 2 for Re = 0) and show the existence of slowly circulating backflow (eddies) within the alveoli. Fluid particles within the central duct do not appreciably enter the alveoli, but flow past them.

Axial mixing coefficients (D^*) in the alveolated ducts are displayed in Fig. 3 as a function of Peclet number ($Pe = \bar{V}d/D$, where $\bar{V}$ is mean flow velocity and d is central duct diameter). Axial mixing is directly influenced by the geometry of the alveolated duct. In particular, the ratio

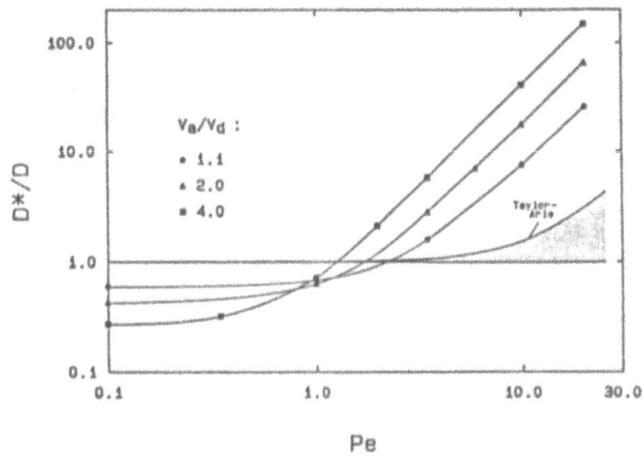

FIGURE 3. Axial mixing coefficients in the alveolated duct model.

of alveolar volume (V_a) to central duct volume (V_d) plays a central role (the curves in Fig. 3 correspond to V_a/V_d of 1, 2 and 4). Values of D^* are markedly different than values within the bounds predicted by straight tube models (stippled region), and the departure from straight tube behavior is proportional to V_a/V_d. At small Pe axial mixing coefficients are less than molecular diffusivity. As Pe increases, flow enhanced dispersion arises and D^* values become much larger than the classical Taylor dispersion result for the straight tubes [11]. Furthermore, flow enhancement in alveolated ducts becomes appreciable at Pe numbers that exhibit negligible flow enhancement in straight tubes.

These results present one example of how microscale geometry (morphology) can impact on macroscale transport coefficients. Other geometric features may be important as well, not only for intrapulmonary mixing [6], but for exchange between the alveolar air phase and pulmonary capillary blood.

3 Alveolar Capillary Exchange

The pulmonary diffusing capacity, D_L, is a lumped conductance specifying the O_2 flow across the alveolo-capillary membrane into blood per unit difference in O_2 tension driving the exchange process. As such D_L accounts for the complex processes of transcapillary diffusion into flowing blood, and chemical reaction within the red cells. Mathematical models of these processes usually assume capillary blood is a homogeneous medium for oxygen uptake. Most oxygen is taken up by red cells, however, and the microscale geometry arising from the particulate (suspension) nature of blood assures that the sink for oxygen in pulmonary capillaries will be discrete rather than continuous.

We looked at oxygen uptake in pulmonary capillaries using a model of particulate blood flow in pulmonary capillaries [5]. The model consisted of red cells as hemoglobin-containing spheres moving in single file suspension through a cylindrical capillary (Fig. 4). Interposed between the capillary

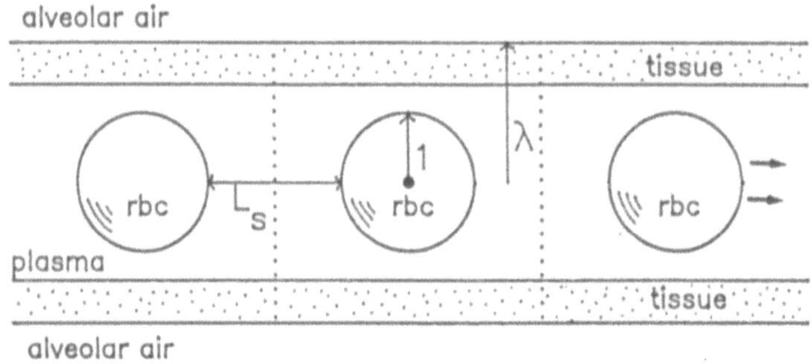

FIGURE 4. Model of two-phase blood flow through capillary tissue unit.

and alveolar gas was alveolar tissue accounting for endothelium, interstitium, epithelium, and surface lining layers. In a reference frame moving with the red cells, the transport equations for free oxygen and bound oxygen within the red cells [3] are, respectively,

$$\partial P/\partial t = D_{O_2}\nabla^2 P - \alpha^{-1}T \quad (3)$$

$$\partial S/\partial t = D_{Hb}\nabla^2 S + C_{tot}T \quad (4)$$

where P is oxygen tension, S is hemoglobin saturation, D_{O_2} and D_{Hb} are oxygen and hemoglobin diffusivities respectively, α is oxygen solubility in the red cell, C_{tot} is heme concentration within the red cell, and T is the net rate of the oxygen-hemoglobin reaction [3]. In the plasma and tissue region, the transport equation for free oxygen is similar to Eq. 3 with coefficients appropriate to these regions and no reaction term; any convective effects in the frame of reference moving with the red cells is assumed negligible [1]. The governing equations were solved by numerical methods [5] to predict the oxygen flow from the alveolar space at a specified oxygen tension into the flowing red cell suspension. Diffusing capacities were computed based on the oxygen flow from alveolar air to red cells, and on the oxygen tension drops driving the exchange process. The overall diffusing capacity, D_L, was resolved into its serial components: D_M, the "membrane" diffusing capacity accounting for all conductance *outside* the erythrocytes, and D_e, the intra-erythrocyte conductance.

Overall diffusing capacity, D_L, decreases as spacing, L_s, between cells increases (Fig. 5, results are normalized and shown for several values of

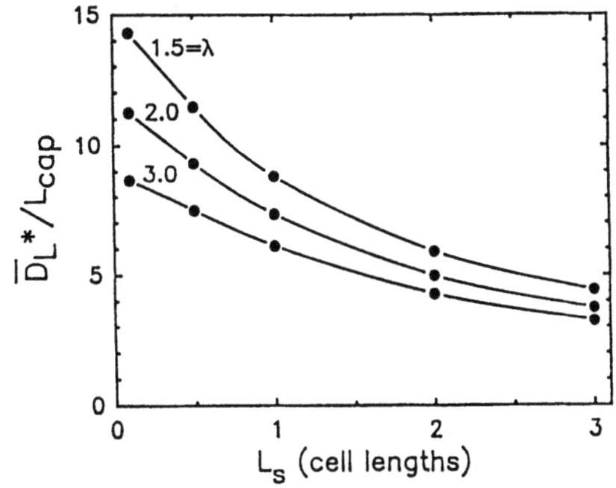

FIGURE 5. Normalized overall pulmonary diffusing capacity.

the capillary tissue size to red cell size, λ). Since the hematocrit of the capillary blood decreases as cell spacing increases, a possible explanation for the reduction in D_L with increasing cell spacing is that D_e decreases as the hematocrit decreases. However, the computations indicate that D_e is a smaller component of D_L than is D_M. Accordingly, the reduction in overall diffusing capacity arises predominantly from a reduction in membrane diffusing capacity, D_M, with increasing red cell spacing, as indicated in Fig. 6

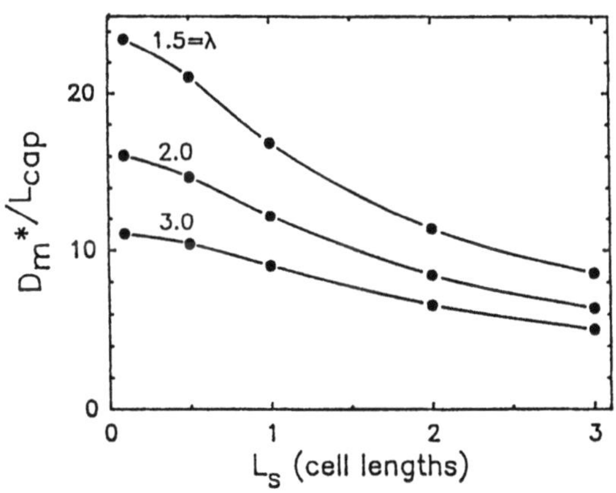

FIGURE 6. Normalized membrane diffusing capacity.

Membrane diffusing capacity decreases with increasing cell spacing because of the way in which oxygen flows across the alveolo-capillary membrane into the discrete oxygen sinks. At any instant oxygen flows predominantly across portions of the membrane adjacent to red cells, while membrane area between red cells participates less (at any instant) in the exchange process. Thus, only a portion of the membrane is functional (at any instant) in the exchange process, and as a result, the membrane diffusing capacity is attenuated. These results imply that D_M is not strictly a morphological parameter [13], but has functional determinants related to capillary hematocrit as well [7].

4 Future Directions

Transport and exchange within the pulmonary acinus are not strictly serial, distinct processes at the microscale level. Accordingly, macroscale conductances for intrapulmonary mixing or alveolar capillary exchange may each be influenced by both local airway transport and exchange with blood. Recent work in dispersion theory [10] indicates that the value of the axial mixing coefficient (and the effective species convection velocity) in conduits can be directly altered by exchange occurring at the vessel walls, and conversely, the effective overall exchange coefficient can be directly altered by convection and diffusion within the conduit. Though the latter phenomenon has been addressed to some extent in the pulmonary system [4], the former has not. The degree to which either phenomena may be manifest in overall respiratory transport awaits further analysis.

Acknowledgements: This work was supported by NIH Grants HL-37106 and HL-33009.

References

[1] Aroesty, J. and Gross, J.F. Convection and diffusion in the microcirculation. *Microvasc. Res.* 2: 247-267, 1970.

[2] Brenner, H. Dispersion resulting from flow through spatially periodic porous medium. *Phil. Trans. R. Soc. Lond.*, 297:81-133, 1980.

[3] Clark, A. Jr., Federspiel, W.J., Clark, P.A.A., and Cokelet G.R. Oxygen delivery from red cells. *Biophys. J.* 47: 171-181, 1985.

[4] Davidson, M.R. and Fitz-Gerald, J.M. Transport of O_2 along a model pathway through the respiratory region of the lung. *Bull. Math. Biol.* 36:275-303, 1974.

[5] Federspiel, W.J. Pulmonary diffusing capacity: implications of two-phase blood flow in capillaries. *Resp. Physiol.* 77: 119-134, 1989.

[6] Federspiel, W.J. and Fredberg, J.J. Axial dispersion in respiratory bronchioles and alveolar ducts. *J. Appl. Physiol.* 64(6): 2614-2621,

1988.

[7] Jouasset-Strieder, D., J.M. Cahill, J.J. Byrne, and Gaensler E.A. Pulmonary diffusing capacity and capillary blood volume in normal and anemic dogs. *J. Appl. Physiol.* 20: 113-116, 1965.

[8] Paiva, M. Theoretical studies of gas mixing in the lung. In: *Gas Mixing and Distribution in the Lung.* edited by L.A. Engel and M. Paiva. New York, NY, Marcel Dekker, pp. 221-286, 1985.

[9] Scheid, P. and Piiper, J. Intrapulmonary gas mixing and stratification. In: *Pulmonary Gas Exchange.* Vol I, edited by J.B. West. New York, NY, Academic Press, pp. 88-131, 1980.

[10] Shapiro, M. and Brenner, H. Dispersion of a chemically reactive solute in a spatially periodic model of a porous medium. *Chem. Eng. Sci.* 43(3): 551-571, 1988.

[11] Taylor, G. Dispersion of soluble matter in solvent flowing slowly through a tube. *Proc. R. Soc. Lond. (A)* 219:186-203, 1953.

[12] Thompson, J.F., Warsi, Z.U.A., and Mastin, C.W. *Numerical Grid Generation.* NY:North-Holland 483pp, 1985.

[13] Weibel, E.R. Morphometric estimation of pulmonary diffusing capacity. I. Model and Method. *Respir. Physiol.* 11: 54-75, 1970/71.

Flow Dynamics of the Nasal Passage

Kevin J. Sullivan and H.K. Chang
Department of Biomedical Engineering, University of Southern California
Los Angeles, California 90089- 1451

1 Introduction

The nasal passage is the natural conduit for directing air into the lungs during breathing at rest and is largely responsible for filtering, warming and humidifying the inspired air. Filtration is achieved primarily by the inertial impaction of airborne particles within the external nose and nasal cavities where the inspired air enters at high velocity and is diverted by abrupt changes in airway direction and geometry [1]. Warming and humidification are accomplished largely by convection, conduction and mass transport within channels formed by the turbinates of the nasal cavities (Figure 1). The large wall surface area and narrow width of these channels increase the transport efficiency [2].

In addition to its capacity to adjust and condition the inspired air, the nasal passage is also the largest single source of flow resistance in the respiratory system [3]. This fact is perhaps not surprising since the geometric features of the nasal passage that are responsible for filtration and heat transfer also utilize and dissipate much of the internal energy of the air stream. Because of its substantial contribution to the total flow resistance of the respiratory system, however, the nasal passage exerts a profound influence on the work of breathing, on the pattern of ventilation and on the mechanical properties of the lungs. The simple, often unconscious act of switching to mouth breathing with exercise, for example, may be provoked by the relatively greater work of breathing that is necessary to maintain nasal breathing at high rates of ventilation. Abnormally high levels of nasal resistance that results from chronic infections, injury or congenital defects of the nasal passage are often associated with chronic lung infections and with disturbances of pulmonary mechanical function. In many instances, alleviating or reducing the nasal dysfunction also improves pulmonary function [4]. Whereas the influence of nasal resistance on normal lung function was first recognized nearly à century ago [5], the physiological processes that link the two have been difficult to identify and remain obscure. This impasse in our understanding of the role of nasal resistance in respiration arises in part from a lack of basic information concerning the fluid dynamic processes that determine nasal flow resistance.

To study nasal function, clinicians rely upon well established rhinomanometric techniques to obtain simultaneous measurements of the pressure drop across and flow through the nasal passage. The resultant trans-nasal pressure-flow relations are often characterized in terms of a single value of resistance that is expressed as the simple ratio of the pressure drop to the flow rate. The information provided by this approach is limited by the fact that the pressure drop across the nasal passage is non-linearly related to the flow, and hence the resistance is more appropriately expressed as a function of the flow rate. A commonly employed alternative is to fit the trans-nasal pressure-flow relation with a second-order polynomial and use the resultant parameters to characterize the flow dependence of nasal resistance. Although an excellent fit to the data is usually obtained by this method, the physical significance of the polynomial relation is unclear and the method offers little insight into the nature of the underlying flow dynamics. In fact, the trans-nasal pressure flow relation is sometimes better fit with a third-order polynomial [6], suggesting that the parameters obtained by this method of analysis should be viewed simply as empirical coefficients that relate pressure with flow.

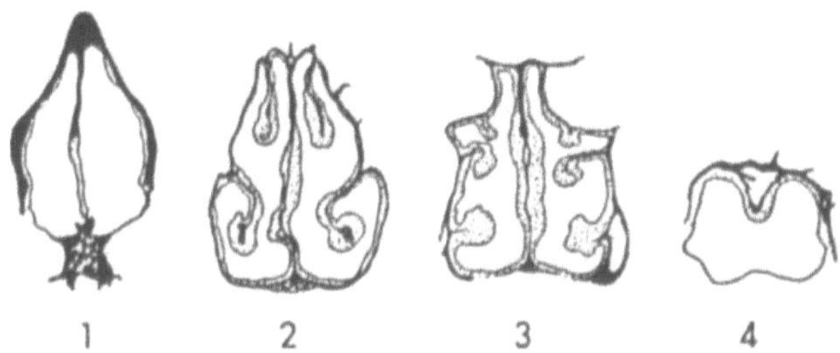

Figure 1. Coronal sections of the nasal cavities illustrating the variation in flow cross-section between the anterior entrance (1) and the posterior apertures or choanae where the air is funnelled into the nasopharynx (4). Bony processes known as turbinates emerge within the middle sections (2,3) of the nasal cavities and divert the flow into channels.

One alternative to the traditional methods of analysis is to assume that the nasal passage is a rigid conduit of fixed internal geometry. In this case, the pressure drop associated with frictional losses within the nasal passage will be influenced by the same factors that determine frictional losses of within, say, a pipe bend. For steady, incompressible flow[1], these are primarily the average velocity of the gas, the gas density and viscosity, and the internal geometry of the conduit. By analogy with pipe flow, the relevant fluid properties and can gathered into the Reynolds number, Re. The principal advantage of this representation is that the frictional losses may be expressed as a function only of Re and are

[1] At rest, the Mach number of flow through the *ostium internum* of the external nose is approximately 0.02

independent of the physical properties of the gas. Provided the nasal passage acts very nearly as rigid conduit, one may expect that the correlation between the pressure loss and the Re of the flow will be similar for different gasses.

The nasal passage is unlike a rigid conduit in many aspects, however. Some regions of the nasal passage, notably those within the external nose and nasopharynx, are flexible and may collapse partially during certain types of breathing. The turbinates of the internal nose are covered with a specialized vascular and erectile tissues that swell in size in response to a variety of environmental and physiological stimuli and thereby alter the internal geometry of the nasal passage and the characteristics of nasal flow. Furthermore, the mucus layer that lines the surfaces of these tissues is deformed and propelled by shear forces that develop at the air-mucus interface [7]. The effect of mucus on nasal flow resistance depends, in part, on its rheological properties which varies substantially both in health and disease. Altogether, these and other characteristics of the nasal passage will contribute to the energy dissipated in the flowing stream but do so through dynamic processes that, *a priori*, are not dependent solely upon the Re of the flow. Determining the relationship between pressure losses and the Re of nasal flow therefore, is a basic step towards identifying the predominant fluid dynamic processes that underlie nasal flow resistance.

2 Pressure-Flow Relations for Steady Flow

The trans-nasal pressure flow relations for near steady flows were determined in trained individuals by drawing test gasses through the nasal passage and through a cylindrical mouthpiece during the performance of a Valsalva maneuver. To avoid the possible reflex changes in the nasal mucosa from prolonged exposure to steady flows, the time required to obtain these data was shortened by slowly increasing the flow of gas through the nasal passage while the subject performed the Valsalva maneuver. Provided that the rate of change of the imposed flow was small, (i.e. $S_t<1$), the instantaneous flow was considered quasi-steady. Using this approach, the pressure-flow relations in one direction and for flow rates between 0.1 and 1.0 L/s were obtained within several seconds.

The dimensionless quasi-steady pressure-flow relations of the nasal passages from two healthy adults are shown in Figure 2. To construct these plots, Re and the dimensionless pressure drop were determined using the dimensions of the mouthpiece (i.d = 2.25 cm) which was comparable in diameter to the adult trachea. In this way, these results may be compared directly with the results of similar measurements made of the pulmonary airways [9,10] and in which the diameter of the trachea was used as a characteristic dimension. A striking feature of these graphs is the fact that a single curve is formed by the data of three different gasses, suggesting that pressure losses within the nasal passages are largely determined by the Re of the flow. A possible exception occurs for Re greater than about 3000 where the dimensionless pressure drop becomes nearly independent of Re and may reflect the influence of other factors, including the partial collapse of the *ostium internum* of the external nose. The magnitudes of air flows associated Re> 3000 , however, are beyond the flow rates normally observed during breathing at rest.

For Re<3000, the slopes of the curves vary continuously. By analogy with pipe flow, this behavior suggests that nasal flow is in transition between purely laminar and purely turbulent flow and that the principal determinants of energy dissipation appears to be both laminar and turbulent friction. For low values of Re, the slopes of the curves approach -1 suggesting the emergence of predominantly laminar flow characteristics within the nasal passage. The practical significance of these results is that for steady flows of the magnitude observed during breathing at rest (typically about 500 ml/s from [8]) the nasal passage behaves as a rigid conduit (or pipe) in which the flow resistance is primarily determined by fluid friction rather than the interaction of the flow stream with deformable structures or with the mucus lining.

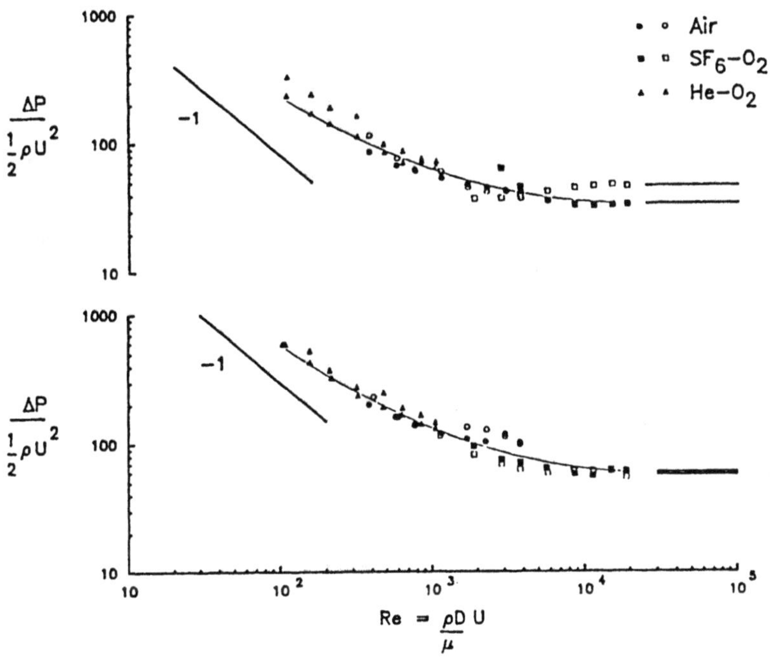

Figure 2. Moody representation of the quasi-steady pressure-flow relations within the nasal passages of two healthy adults. Closed symbols represent inspiratory flows, open symbols represent expiratory flows. Gas mixtures of helium and sulfur-hexaflouride contain 20 % oxygen.

The correlation between the dimensionless pressure drop and the Re of nasal flow is similar to that obtained for flow through the pulmonary airways. Figure 3 compares the average results from 3 subjects with the dimensionless relations obtained for the first 4-5 generations of the pulmonary airways (excluding the larynx) from [9] and for all of the airways between the mouth and alveoli from [10]. Although fluid friction within the central airways is a major source of pressure loss in the lungs, its contribution appears small in comparison to frictional losses in the nasal passage. A significant component of the static or lateral pressure in these airways, therefore, results from the energy required to drive flow through the nasal passage. Inasmuch as the lateral pressure affects the diameter of these airways, the flow resistance of the central airways may be directly influenced by the resistance of the nasal passage. This link between nasal and central airway flow dynamics may help to explain the increase in pulmonary flow resistance that is observed to occur in response to an increase in nasal resistance [5,12].

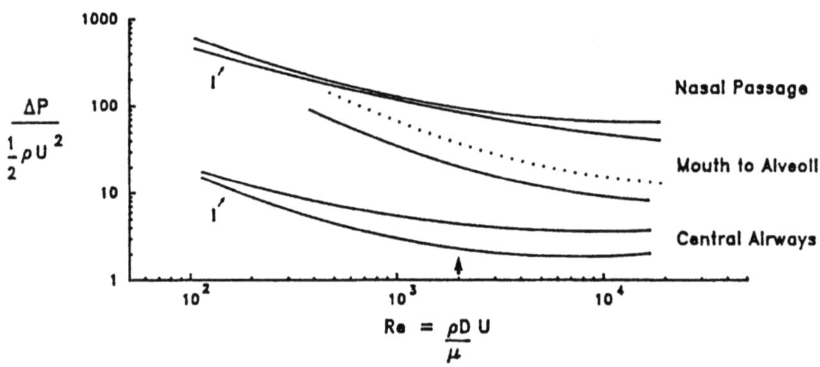

Figure 3. Dimensionless pressure losses within the nasal passages, central airways and lungs. The arrow defines the upper limit of Re for breathing at rest. The dotted line represents the data from [10] adjusted for mouthpiece diameter. The letter I identifies the curves obtained for inspiratory flows

The dimensionless pressure drop across the airways between the mouth and alveoli includes the pressure drop across the larynx. In this case, the total frictional losses within the pulmonary airways are about the same order of magnitude as those within the nasal passages, thus confirming published observations that nasal and total pulmonary resistances are comparable [3]. Because the slopes of these curves vary continuously with Re, there exists no unique values of nasal or pulmonary flow resistance in the range of flow associated with breathing at rest. The coincident changes in nasal and pulmonary flow resistance with Re, however, raises the interesting possibility that nasal resistance roughly forms a constant proportion of total pulmonary resistance over a wide range of Re.

3 Pressure-Flow Relations for Unsteady Flow

Although the Re appears sufficient to describe the dimensionless pressure drop for quasi-steady nasal flows, temporal variations in nasal flow occur during normal respiration and flow steadiness cannot be assumed. To determine the characteristics of the nasal resistance to unsteady flows, we examined the trans-nasal pressure flow relations obtained using flow oscillations of constant amplitude and at frequencies between 1 and 16 Hz. This range of frequency spans the harmonic content of the respiratory flow waveforms that are produced during breathing at rest and voluntary hyperventilation. At all frequencies examined, the pressure-flow relations formed a loop or hysteresis, thus revealing components in the resultant pressure that are out of phase with the imposed flow. To identify these components, we noted that the energy contained in the harmonics that result from non-linear relation between pressure and flow were small in comparison to the energy found in the fundamental frequency of the oscillation. As a consequence, for a flow oscillation of amplitude, A, and frequency, ω, the trans-nasal pressure-flow relations may be characterized by the describing function,

$$N(A,\omega) = n_p(A,\omega) + jn_q(A,\omega)$$

where $n_p(A,\omega)$ is the linearized component of the pressure that is in phase with the flow (i.e. the linear resistance), $n_q(A,\omega)$ is the linearized component in quadrature with flow and $j = \sqrt{-1}$. The describing function N (A,ω) is analogous to the transfer function of a linear system with the exception that its value is determined by the amplitude of the oscillation as well as the frequency [12].

The results obtained from 3 subjects at 1 and 16 Hz are plotted against the Re of peak flow in Figure 4. The linearized flow resistance, n_p, at 1 Hz. is similar to the steady flow resistance over almost the entire range of Re, suggesting that flow unsteadiness of the magnitude found in these oscillations has a negligible effect upon the overall nasal resistance. A virtually identical result is found concerning the flow resistance of the central airways for frequencies up to 2 Hz [9]. Whereas n_p tends to increase with increasing Re at both 1 and 16 Hz, the increase is greater at 1 Hz as Re increases above roughly 3000. The relatively abrupt increase in the nasal resistance to oscillatory flows in this range of Re can be expected to greatly increase the work of breathing through the nose and thus may provide a stimulus to switch to mouth breathing with exercise hyperpnea. On the other hand, the fact that n_p measured at 16 Hz is less than the corresponding 1 Hz values for Re>3000, indicates that n_p becomes frequency dependent with the increasing amplitude of peak nasal flow.

Frequency dependent changes in the resistance to flow through the complex and bifurcating geometry of the nasal passage may result from two fundamental processes: 1) the re-distribution of flow between parallel pathways and 2) the distortion of the velocity profiles by inertia. An examination of Figure 4, reveals that the in-quadrature component, n_q, for flow oscillations at 1 Hz becomes negative at high levels of Re, indicating that flow leads pressure over most of the cycle. By analogy with the purely capacitive reactance of a linear system, this behavior is consistent with the expansion of compliant regions of the nasal passage and oral cavity, particularly at the high lateral pressures associated with this range of Re. At 16 Hz, however, n_q is positive, thus implying that the principal in-quadrature component arises from the inertial forces related to the time dependent acceleration of gas. Under certain circumstances, the shunt or parallel pathway for flow formed by the distensible airways causes n_p to decrease with increasing frequency [13]. In contrast, distortion of the velocity profiles by fluid inertia may be expected to increase resistance with increasing frequency [9]. Whereas both processes will contribute to the global pressure-flow relations for unsteady flows, the large negative component of n_q together with the pattern of change in n_p suggests that nasal resistance to oscillatory flows is predominantly influenced by flow re-distribution for Re>3000.

In summary, the trans-nasal pressure-flow relations in healthy adults appears to be uniquely determined by the Re of nasal flow under condition of near-steady flow. For unsteady flows associated with simple harmonic flow oscillations between 1 and 16 Hz, nasal flow resistance is comparable to steady flow resistance for Re<roughly 3000. These results suggests that nasal flow resistance during breathing at rest is largely determined by frictional losses relating to convective acceleration and viscous forces within the nasal passage.

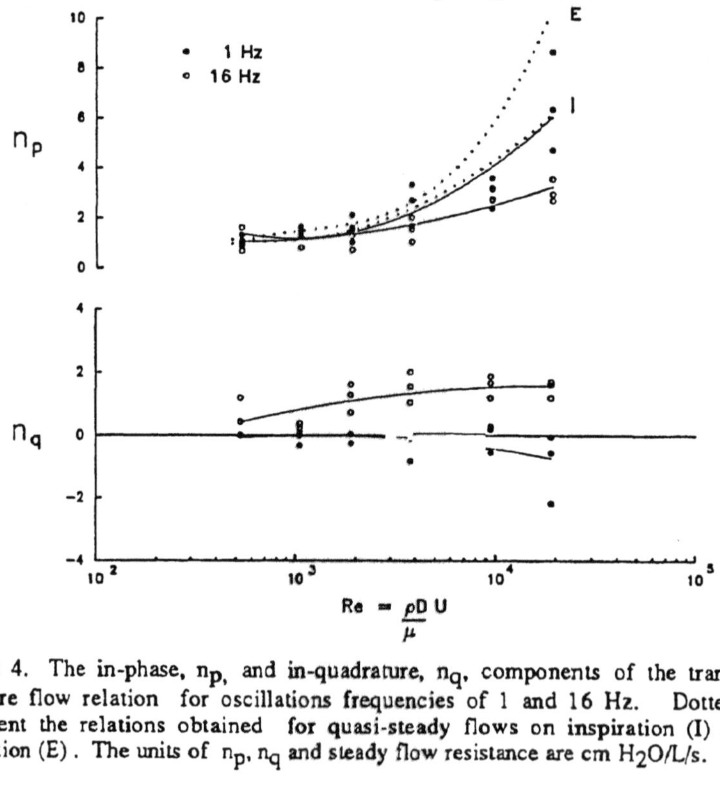

Figure 4. The in-phase, n_p, and in-quadrature, n_q, components of the trans-nasal pressure flow relation for oscillations frequencies of 1 and 16 Hz. Dotted lines represent the relations obtained for quasi-steady flows on inspiration (I) and on expiration (E). The units of n_p, n_q and steady flow resistance are cm H_2O/L/s.

References

[1] Landahl, H.D., and S. Black. Penetration of airborne particulates through the human nose. J. Ind. Hyg. Toxicol. 29: 269-277, 1947.

[2] Ingelstedt and Toremalm. Airflow patterns and heat transfer within the respiratory tract. Acta. Physiol. Scand. 51: 1-14, 1961.

[3] Ferris, B.J., J. Mead and L.H. Opie. Partitioning of respiratory flow resistance in man. J. Appl. Physiol. 19: 653-658, 1964.

[4] Ogura, J.H., K. Towage, R. Dammkoehlet, J.R. Nelson, and M. Kawasaki. Nasal obstruction and the mechanics of breathing. Arch. Otolaryngol. 83: 135-150, 1966.

[5] Rohrer, F. Der Stromungswiderstand in den menchlichen Atemwegen und der einfluss der unregelmaggigen Verrzweigung des Bronchialsystems aus den Atmungsverlauf in verschiedenen Lungenberzirken. Arch. Ges. Physiol. 162: 225-300, 1915.

[6] Schumacher, M.J., J.A. Gaines, and B. Bescript. Computer aided rhinometry: analysis of inspiratory and expiratory nasal pressure-flow curves in subjects with rhinitis. Comput. Biol. Med. 15: 187-195, 1985.

[7] Kim, C., C.R. Rodrigues, M.A. Eldridge, and M.A. Sackner. Criteria for mucos transport in the airways by two phase gas-liquid flow mechanism. J. Appl. Physiol. 60: 901-907, 1986.

[8] Swift, D.L., and D.P. Procotor. Access of air to the respiratory tract. In: Brain, D., Proctor, D.F., Reid L.M., eds. *Respiratory Defense Mechanisms*. New York: Marcell Dekker Inc., pp 95-124, 1977.

[9] Isabey, D. and H.K. Chang. Steady and unsteady pressure flow relations in central airways. J. Appl. Physiol. 51: 1338-1348, 1981.

[10] Jaffrin, M.Y., and P. Kesic. Airway resistance: a fluid mechanical approach. J. Appl. Physiol. 36: 354-361, 1974.

[11] Ogura, J.H. and J.E. Harvey. Nasopulmonary mechanics. Experimental evidence of the influence of the upper airway upon the lower. Acta. Otolaryngol. 71: 123-132, 1971.

[12] Gelbe, A., and W.E. Vander-Velde. *Multiple input describing function and nonlinear system design*. New York: McGraw-Hill, Inc., 1968.

[13] Chang, H.K. Flow dynamics in the respiratory tract. In: Chang, H.K., and Paiva, M., eds: *Respiratory Physiology: an analytical approach*. New York: Marcell Dekker Inc., pp 57- 138, 1989.

PULMONARY CIRCULATION

Elasticity of Pulmonary Blood Vessels in Human Lungs

R. T. Yen, D. Tai*, Z. Rong** and B. Zhang**, Dept. of Biomedical Engineering, Memphis State Univ., Memphis, TN 38152, *Dept. of Radiology, Univ. of Tenn., Memphis, TN 38163, ** Shanghai Medical Univ., P. R. of China

Introduction

The mechanical properties of pulmonary blood vessels are essential factors influencing the distribution of pulmonary blood flow, regional volume distribution of blood and pulse wave propagation throughout the lung. These properties affect the pressure-flow relationship, and the change in total blood volume in response to an alteration in blood pressure. In order to formulate such problems, mechanical property data for blood vessels of all generations of the human lung are needed and are gathered in our laboratory. In particular, vessel diameter as a function of transmural pressure are measured for each vessel generation.

Although extensive blood vessel elasticity data exists such as dog, cat and rabbit, no data for the elasticity of the blood vessels of the human lung is presently available. This chapter reports our research of the mechanical properties of both arteries and veins of the human lung. The microvascular elasticity data have been reported in our earlier paper (1988).

Postmortem Human Lungs

Since there is no way to obtain a complete set of human data on living persons, the use of postmortem material is the logical answer. For the following observations we believe that the measurements obtained from postmortem lungs are valid and can be properly used. (a) Previous unpublished work by us has shown that successively removed samples of cat lung (taken antemortem and up to a period of 48 hours postmortem from the same refrigerated animal) show no changes by light microscopy except for increased translucency. (b) Hildebrandt and co-workers (1971) found no difference between human autopsy material up to 36 hours postmortem and surgical specimens in tension-length studies of parenchymal strips of human lungs. (c) Weibel, Bachofen and Roos (1975) have found no changes by electron microscopy in the normal human lung up to eight hours postmortem.

Lungs are removed from individuals who die from unknown or accidental causes and in whom there is neither lung injury nor gross evidence of lung pathology at the time of autopsy, nor reason to suspect lung pathology as a cause of death. Individuals with gross evidence of cardiac disease are excluded. The time between death and autopsy as a general rule does not exceed 24 hours. The specimen information is listed in Table 1.

Table 1 Specimen Information

Age yr.	Sex	Hours Post-Mortem	Preparation	Perfusion Direction
41	M	16	Right Lung	Antegrade
35	M	4	Right Lung	Retrograde
20	M	6	Right Lung	Antegrade

Method

Human lungs of good structural integrity and with no apparent edema are obtained at autopsy within eight hours of time of death. After cannulation, each lung is placed in the lucite box, which is maintained at 37^{0}C and appropriately connected to the perfusion reservoir, vacuum source, etc. The blood vessels are first purged with blood by flushing with normal saline containing 6% dextran. They are then perfused with 30% barium sulfate ($BaSO_4$) and 5% methyl cellulose in glucose solution. After full expansion with sufficiently high pleural pressure, the lung is then preconditioned by inflating and deflating ten times by varying the pleural pressure from 0 to 20 cm H_2O. Preconditioning is a well recognized procedure in mechanical testing in order to stabilize the mechanical properties of living tissues; it also ensures that all alveoli are open and the entire lung is perfused properly. Following preconditioning, the lung is perfused at a specified pressure. Then the left atrium is closed off to allow the pressure to equilibrate in all blood vessels in the no-flow condition. During perfusion at a given pressure, a constant pleural pressure of -10 cm H_2O is maintained. This fixed pleural pressure ensures a fixed lung inflation so that the location of each blood vessel with respect to the reference wire on the box does not change as the perfusion pressure is changed. This allows the identification of vessels on successive x-ray films taken at different perfusion pressures. At each perfusion pressure, a chosen region on the lung is exposed to x-ray and photographed with Kodak Blue Brand film. Because the average barium sulfate particle size is 15 μm, it cannot pass through the capillary blood vessels, thus insuring that only the arterial vessels or venular vessels appear in the radiographs. By measuring the vessel size at different perfusion pressure, its distensibility can than be quantified.

Results

Results for the x-ray photography study of pulmonary blood vessels are grouped with diameters in different ranges. After all measured values of vessel diameters of human lung are corrected for penumbra and scaled to their true dimensions, they are classified into several ranges based on their diameter at near zero transmural, D_0. The results are presented in the form of percentage change in the diameter D as a function of the p_a-p_{PL} in their respective diameter ranges. An almost linear relationship between percentage diameter changes and transmural pressure is usually obtained within a pressure range of 50 cm H_2O, because of the tethering of the lung interalveolar septa on the vessels. The regression may be written as $D/D_0=1+\alpha(p_a-p_{PL})$, where α is the compliance coefficient. Fig 1 shows the results for the pulmonary arteries and Fig 2 shows the results for the pulmonary veins.

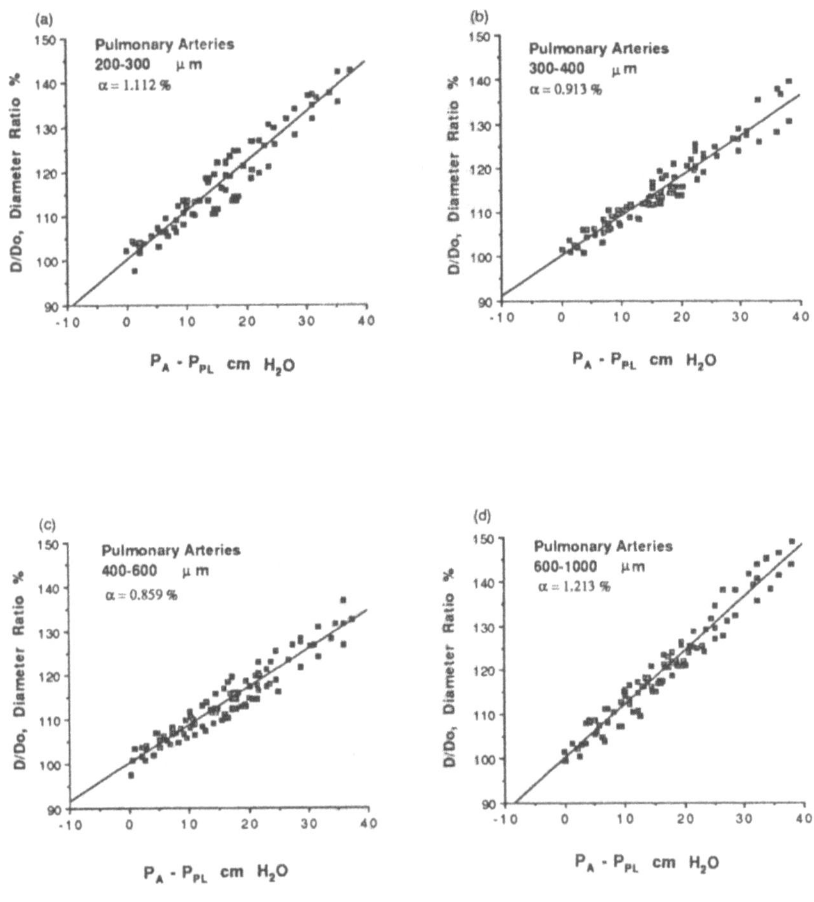

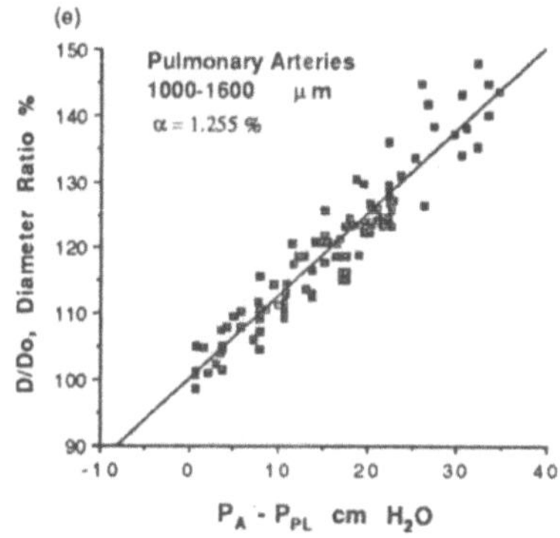

Figure 1. Percentage changes in diameter as a function of transmural pressure p_a-p_{PL} which is the pressure difference between arterial pressure p_a and pleural pressure p_{PL} for pulmonary arteries of human lung. (a) For vessel size 200-300 μm (b) For vessel size of 300-400 μm (c) For vessel size of 400-600 μm (d) For vessel size of 600-1000 μm (e) For vessel size of 1000-1600 μm. A linear relationship is observed within the pressure range.

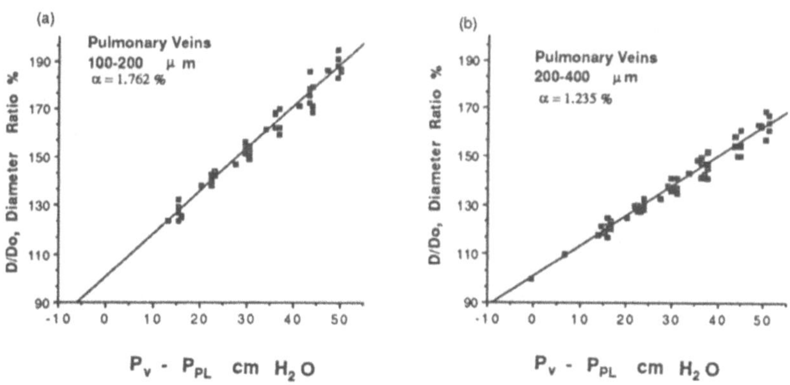

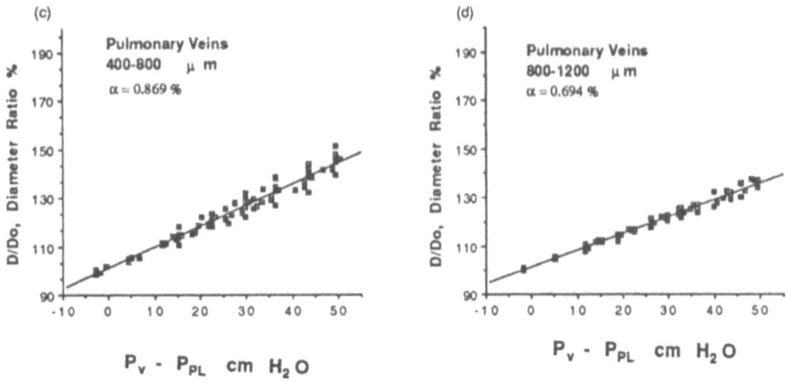

Figure 2. Percentage changes in diameter as a function of transmural pressure p_v-p_{PL} (pressure difference between venous pressure p_v and pleural pressure p_{PL} for pulmonary veins of human lung. (a) For vessel size of 100-200 μm (b) For vessel size of 200-400 μm (c) For vessel size of 400-800 μm (d) For vessel size of 800-1200 μm. The veins do not collapse when the transmural pressure is negative.

Discussion

(a) Nonlinearity of the pressure-diameter relationship

The compliance of a blood vessel as expressed by the percentage changes of diameter vs change in transmural pressure, depends partly on the ratio of the wall thickness to vessel diameter, and partly on the Young's modulus of the wall material. A simple estimation can be based on the assumption of thin walled tube. According to the well known Laplace formula, the tensile stress in the wall of a tube of diameter D is

$$\sigma = (p_a - p_{PL}) \left(\frac{D}{2h}\right) \tag{1}$$

where h is the wall thickness. The strain is σ/E, which is also the change of diameter divided by the diameter. Hence, the compliance is

$$\frac{(change\ of\ dia/D_0)}{(p_a - p_{PL})} = \frac{D}{2h}\frac{1}{E} \qquad (2)$$

Our results show that the pressure-diameter relationship is linear within the pressure range of 50 cm H_2O. According to Eq. (2), the pressure-diameter relationship will be linear if the right hand side of that equation, $D(2hE)^{-1}$, remains constant as the pressure varies. Now, for blood vessel in general, D increases with increasing pressure, h decreases, whereas E increases. That the Young's modulus E increases with stress is well known (1972). The p-d relationship will remain linear only if the changes in D and hE remain exactly proportional. We suspect that this is a consequence of tethering by interalveolar septa. Note also that the linearity is necessarily limited by the range of pressure used. If the pressure increased indefinitely, eventually the increase in E will outstrip the increase of (D/h) and p-d relationship will become nonlinear. Fig. 3 shows the p-d relationship of a vein of 500 μm diameter. The value of the pressure is first decreased from 47 H_2O to a lowest value of 13.7 cm H_2O and increases again to the original value of 47 H_2O by 5 cm H_2O step. The pressure is then further increased to a final value of 92.2 cm H_2O. As can be seen the p-d relationship is linear within the pressure range of 50 cm H_2O, and shows very little hysteresis. Beyond the pressure of 50 cm H_2O, the curve becomes quite nonlinear.

(b) Compliance of the vessels

The overall trend of the human pulmonary arterial tree seems to be such that the large arteries are more compliant and the smaller ones are less so, until a minimum is reached for vessels in the 400-600 μm range. Then the compliance increases again, and the vessels become very compliant at the capillary blood vessel level. Whereas the human venous tree show that smaller veins are more compliant than large veins.

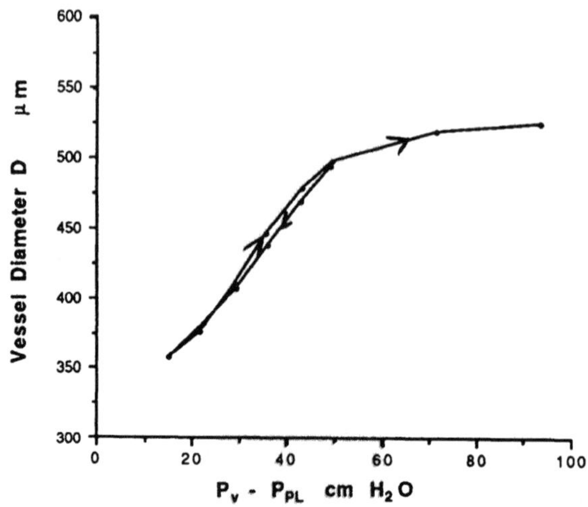

Figure 3. A typical pressure-diameter relationship of a vein of 400-800 µm diameter range. The transmural pressure is first decreased from 47 cm H_2O to a lowest value of 13.7 cm H_2O and then increased again to a highest value of 92.2 cm H_2O.

References

Bachofen M, Weibel ER, Roos B (1975) Postmortem fixation of human lungs for electron microscopy. *Am Rev Respir Dis* IIIi:247-256.

Patel DJ, Vaishnav RN (1972) The rheology of large blood vessels. In Bergel DH (ed) *Cardiovascular Fluid Dynamics*, Academic Press, New York: 22-65.

Sugihara T, Martin CJ, Hildebrant (1971) Length-tension properties of the alveolar wall in man. *J Appl Physiol* 30:874-878.

Yen RT, Sobin SS (1988) Elasticity of arterioles and venules in postmortem human lungs. *J Appl Physiol* 64:611-619.

DISTENSIBILITY OF THE PULMONARY CAPILLARIES

J. S. Lee and L. P. Lee
Dept. of Biomedical Engineering, Univ. of Virginia
Charlottesville, Virginia, 22908 USA

Introduction

A sheet flow has been employed as a morphometric idealization of the vascular space of the pulmonary capillaries (Sobin and Fung, 1972 and Fung, 1980). In this model, blood flows in between two membranes and around the posts holding the membranes apart at a sheet thickness h. Because of the distensibility of the pulmonary capillaries, the thickness increases when the transmural pressure (P_{tm}), the capillary blood pressure minus the alveolar gas pressure, is increased. As one increases the transpulmonary pressure (P_{tp}), the alveolar gas pressure minus the pleural pressure, to inflate the lung to a larger volume, the surface area of all alveolar sheets becomes larger. Since the blood volume of the pulmonary capillaries is related to the product of the sheet thickness and the surface area in the sheet flow model, we have the following incremental relation to describe the distensibility of the pulmonary capillaries

$$\Delta V_c/V_c = \Delta P_{tp}/E_1 + \Delta P_{tm}/E_2 \qquad (1)$$

where E_1 and E_2 are elastic moduli and Δ represents the increment of the quantity that follows.

Based on a recently developed method which measures the density of blood, we convert the density variation to a hematocrit variation and use it to quantify the volumetric change of the pulmonary capillaries. For an in vitro perfused lung, we could independently impose a cyclic change in the transpulmonary or transmural pressure or both. In this paper, we used the density method and the in vitro preparation to assess the relations between the pressure changes and the volumetric change for the determination of the two moduli. This procedure offers an in vitro dynamic measurement of the distensibility in complement to the static measurement made by Fung and Sobin (1972).

In vitro Experimentation

Five rabbits were exsanguinated through the carotid artery. The lung was isolated and connected to a perfusion circuit consisted of a venous reservoir and an arterial reservoir. Blood was pumped from the venous reservoir to the arterial reservoir. Through an overflow arrangement in the latter, the arterial pressure was maintained at a constant level of 18 mmHg. The level of the venous reservoir was adjusted to yield a venous pressure of 6 mmHg. At a port of the venous connection, the blood was withdrawn to flow through the U-tube of the density meter and then back to the venous reservoir. Based on the resonance frequency of the U-tube, the density of the blood was measured and reported in digital form at a rate of 6 samples per second.

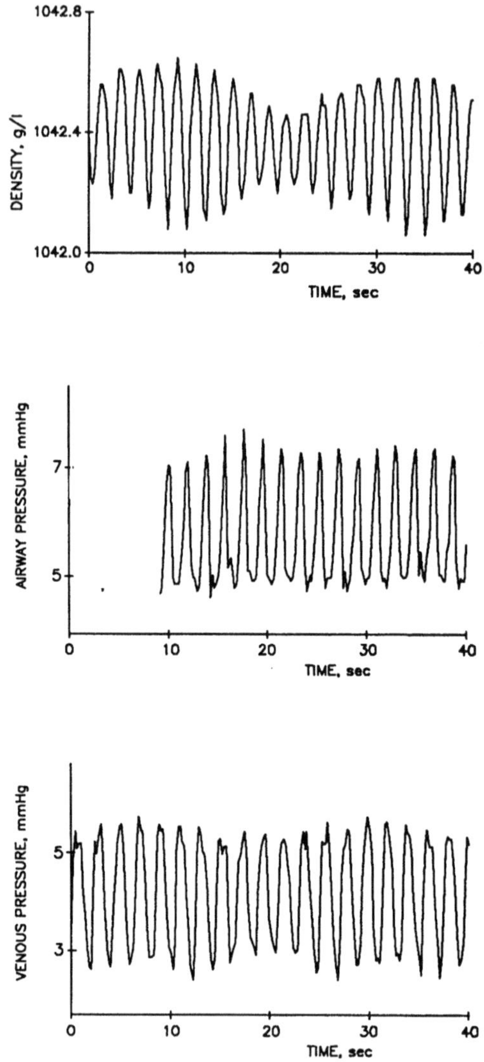

Figure 1. Variation in density of pulmonary venous blood, airway pressure (ω_1 = 30 cpm) and pulmonary venous pressure (ω_2 = 28 cpm)

A 0.5 ml bolus of saline was injected into the arterial catheter for the measurement of the density reduction in the venous blood. With appropriate conversion, the indicator dilution curve or the transfer function characterizing the flow dispersion of the pulmonary vasculature can be derived from the density reduction.

First the lung was ventilated with air containing 5% CO_2 at an end expiratory pressure of 5 mmHg and an inspiratory pressure of 7 mmHg with no time variation in the arterial and venous pressure. (This ventilation was employed throughout the whole experiment in order to maintain the inflation of the lung). Because of the opposite role of the airway pressure in P_{tp} and P_{tm}, the cyclic changes of these two pressure are 180° out of phase. They combine to effect a cyclic change in V_c at the ventilation frequency.

Then we connected the airspace of the arterial and venous reservoir to another respirator to effect a cyclic change in the arterial and venous pressure at a frequency different from that of the ventilation. As a result, another harmonic oscillation was present in P_{tm} to deform the pulmonary capillaries while the mean pressure remained unchanged. The addition of these two oscillations leads to a fast volumetric variation with its amplitude varying at a slower frequency (the difference of the ventilation and perfusion frequency). As an analogy, this is similar to the beating oscillation when two sound waves of different frequency are mixed. Because of the Fahraeus effect, the tube hematocrit of the capillary blood is lower than the discharge hematocrit, we expect the volumetric variation to produce a beating oscillation in the density of blood flowing out of the lung. Such an oscillation is shown in the top panel of Fig. 1. We repeated this experiment with a larger oscillation in the arterial and venous pressure.

Harmonic representations

First we counted the crossings of the tracheal pressure through its mean value to determine the ventilation frequency ω_1. Then we expressed the tracheal pressure as a Fourier series with ω_1 as its fundamental frequency. The net change of the tracheal pressure over one ventilation cycle ΔP_A is taken as twice of the amplitude of the first harmonic of the Fourier series. Similar procedure was used to determine the "perfusion" frequency and the net change in the arterial pressure ΔP_a and that of the venous pressure ΔP_v. For the density, we obtained two net changes $\Delta\rho_1$ for the ventilation frequency and $\Delta\rho_2$ for the perfusion frequency.

Density variation and volumetric change

The red blood cells have a density (ρ_r) about 65 g/l higher than that of plasma (ρ_p). The density of blood (ρ) relates to its hematocrit (H) and the densities of its two components by the following equation:

$$\rho = \rho_p + (\rho_r - \rho_p)H \tag{2}$$

Since the density of plasma and red blood cells are not affected by the changes in the transpulmonary and transmural pressure, we expect

$$\Delta H = \Delta\rho /(\rho_r - \rho_p) \tag{3}$$

When the blood volume of the pulmonary capillaries is reduced, the blood outflowing from the capillaries will exhibit, because of the Fahraeus effect, a transient reduction in the hematocrit (or density). For the case that the volumetric variation is periodic, the hematocrit variation in the blood discharged from the pulmonary capillaries is also periodic with the net change expressed as ΔH_d. As the cyclic variation is carried by the blood to the sampling site, it is attenuated by a factor A^q to form the measured hematocrit variation ΔH, i.e.

$$\Delta H_d = \Delta H/A^q \tag{4}$$

An investigation on the attenuation of a cyclic variation transported by a dispersive blood flow suggests that we can estimate A as the modulus of the Fourier transform of the transfer function. The quotient q represents the fraction of the mean transit time for blood to flow from the pulmonary capillaries to the sampling site over the mean transit time for blood to flow from the pulmonary artery to the sampling site.

Together with the equation developed by Friend and Lee (1990) relating hematocrit variations with volumetric changes, we have

$$\Delta V_c/V_c = \Delta\rho/[W_1(\rho_a - \rho_w)\ \omega\ T_m\ A^q] \tag{5}$$

where $W_1 = \sin(\omega t_c/2)/(\omega t_c/2)$, ρ_a is the arterial blood density, ρ_w the density based on the whole lung hematocrit H_w, T_m is the mean transit time of blood to flow through the pulmonary vasculature, and t_c the mean transit time of red blood cell to traverse the capillaries. When the density variation and frequency associated with the ventilation or perfusion are used in the equation above, we obtain the estimate on the corresponding volumetric change of the pulmonary capillaries.

In equation 5, $\Delta\rho$ is determined from the measured density variation, the attenuation A from the fit of the dilution curve, the mean transit time T_m from the dilution curve. The sheet thickness h is taken as 7.4 μm, t_c as 0.23 T_m, and q as 0.52. The density difference $\rho_a - \rho_w$ is computed from the assumption that H_w is 90% of H_a. The detailed derivation of equation 5 and the value selections are given by Lee (1986), Lee and Lee (1989) and Friend and Lee (1990).

Distensibility

The experiments were carried out at three levels of oscillations in the arterial and venous pressure, namely a ΔP_a of about 0, 3 mmHg, and 6 mmHg. The dependence of ΔV_c on ΔP_a is shown in Fig. 2. Due to the dampening of the oscillating arterial and venous pressure to form the oscillating capillary blood pressure, we take $\Delta P_{tm} = 0.55\ \Delta P_a$ (Gray and Lee, 1990). For this experiment, the airway or alveolar pressure does not contain a harmonic with the perfusion frequency as its fundamental frequency. Accordingly, the results shown in Fig. 2 indicate that the volumetric change of the pulmonary capillary is linearly dependent on the transmural pressure. Calculated from the slopes of the data in Fig. 2, the mean and standard deviation of the elastic moduli E_2 is 81±17 mmHg.

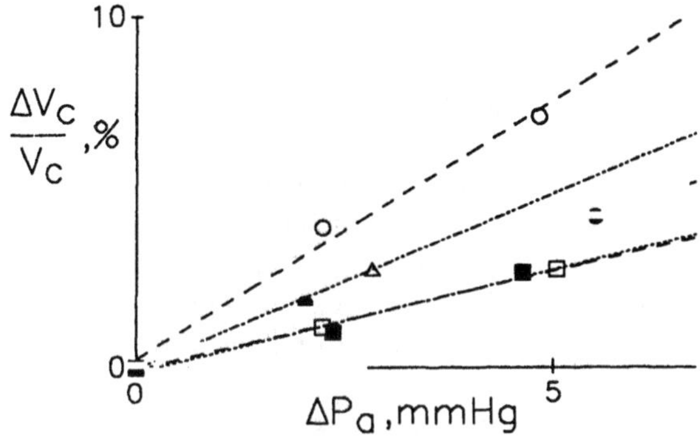

Figure 2. The relation between $\Delta V_c/V_c$ and ΔP_a.

We have shown previously that $\Delta \rho$ is linearly related to ΔP_A. As the ventilation compresses the pulmonary capillaries to generate an outflow, the resistance offered by the pulmonary arterioles and venules causes an change in the capillary pressure. In essence the ventilation effects an oscillation in the capillary pressure with $\Delta P_c = 0.58\ \Delta P_A$ (Gray and Lee, personal communication). Thus, for the case with only the ventilation, we have

$$\Delta V_c/V_c = \Delta P_A/E_1 - 0.42\Delta P_A/E_2 \qquad (6)$$

From the relation between $\Delta V_c/V_c$ and ΔP_A for the experiments with ventilation only and the previous estimate of E_2, we obtain this value 36±8 mmHg for E_1.

Discussion

The ranges of pressure selected for the present experiment are within the physiological ranges. For these ranges, we find that the blood volume of the pulmonary capillaries is affected by the transpulmonary and transmural pressure in a linear fashion. The value of E_2 found by the density method for rabbit's pulmonary capillaries is about three times larger than the value of 25 mmHg reported by Fung and Sobin (1972) from the dimensional changes of polymer casts of cat's pulmonary capillaries. This difference in distensibility could reflect the difference between these two species or that between our dynamic measurement at a frequency of about 28 cycles per minute and their static measurement.

We also find that E_1 is smaller than E_2. It implies that the transpulmonary pressure is more effective in producing a change in the capillary blood volume than the transmural pressure. An analysis on how the geometry of the alveoli could affect the way that the transpulmonary pressure acts to change the blood volume of the pulmonary capillaries needs to be done in order to address the question whether the distensibility of the membrane structure in the direction of the sheet thickness is different from that in the direction parallel to the alveolar surface.

For the larger pulmonary arterioles and venules, the Fahraeus effect is less than that for the pulmonary capillaries. Because their blood volume is also much smaller than that of the capillaries, their deformation leads to a much smaller contribution to the density variation. Consequently the present analysis regarding the density variation to originate from the pulmonary capillaries is reasonable. Although the pulmonary vasculature has been considered as a parallel network, the attenuation analysis based on the measured transfer function could account for the dispersion resulting from a small irregular flow distribution among the pulmonary capillaries.

In summary, the methodology presented here provides a direct measurement on the distensibility of the pulmonary capillaries. As we need the ventilation to maintain the inflation of the lung, the use of two frequencies allows us to separate the contribution from the transpulmonary and transmural pressure in order to assess their effect on the deformation of the pulmonary capillaries. These dynamic measurements could be extended to quantify the viscoelastic properties of the pulmonary capillaries.

ACKNOWLEDGEMENTS This research is supported by grants HL 36285 and HL 40893 National Institute of Heart, Lung, and Blood.

REFERENCES

Friend M, Lee JS (1989) Red blood cell motion and hematocrit distribution in a deforming capillary, *J of Biomechanical Engineering*, in press.

Fung YC (1980) *Biodynamics: Circulation*, Springer-Verlag, New York.

Fung YC, Sobin SS (1972) Elasticity of the pulmonary alveolar sheet, *Circ Res*, 30:451-469.

Gray R, and Lee JS (1990) Dynamic pressure flow relationship of the pulmonary circulation, *FASEB J*, 4:A1107.

Lee JS (1986) Microvascular hematocrit of the lung, In Schmid-Schonbeim G, Wu S, and Zweifach BW (eds) *Frontiers in Biomechanics*, Springer-Verlag, New York, p. 353-364.

Lee JS, Lee LP (1989) Volumetric change of pulmonary capillaries as assessed by a density method, In Lee JS and Skalak TC (eds) *Microvascular Mechanics*, Springer-Verlag, New York, p.147-162.

RECRUITMENT OF PULMONARY CAPILLARIES

Wiltz W. Wagner, Jr., Depts. of Anesthesiology and Physiology-Biophysics, Indiana University Medical School, Indianapolis, IN

Recruitment of pulmonary capillaries is an important component of gas exchange reserve that is utilized during exercise to meet the demand for increased oxygen uptake. However, important gaps exist in our understanding of the way in which alterations of pulmonary hemodynamics, with the attendant redistribution of pulmonary blood flow, affect capillary recruitment. The major reason that many aspects of pulmonary microcirculatory function are poorly understood is the considerable technical difficulty in studying pulmonary microvessels directly. The classical direct approach of in vivo microscopy is plagued by problems with tissue movement during the cardiorespiratory cycles. Nevertheless, important information has come historically from in vivo microscopy of the lung. The directness of this approach continues to make the technique attractive today.

In the early 1960s, Giles Filley instigated work in Denver on the pulmonary microcirculation using in vivo microscopy. He and I had the good fortune of discovering a location on the surface of the lung under the second rib that moved only slightly with respiration (Wagner and Filley, 1965). A transparent window was inserted into the chest wall over this area and the chest was closed to form an airtight seal. Later a suction manifold was added to the bottom of the window frame so that movement was reduced to <10 μm with each breath (Wagner, 1969). This permitted us to make observations of the same field over many hours and thereby to use the same arterioles, capillaries, and venules as their own controls for hemodynamic studies. The preparation gave us the ability to observe directly the pulmonary gas exchange vessels in a still field using high magnification in an animal with normal blood gases, cardiac output, blood pressures, and, as best we could determine, an undamaged microvasculature. This enabled us to make a number of studies of how the pulmonary capillaries responded to changes in pressure and flow. The majority of our observations have been made on the uppermost surface of the lung. This location is an interesting part of the pulmonary circulation, because that region is where much of the gas exchange reserve lies in terms of capillaries that can be recruited.

In our early studies we noticed that red cells consistently perfused more capillaries when the animals inspired hypoxic gas mixtures (Wagner and Latham, 1975). To quantify this recruitment, we made drawings of all capillaries that were perfused by red blood cells in a given field during normoxia and hypoxia. By measuring the total length of the perfused capillaries per unit field area, we obtained an index of capillary recruitment. This capillary recruitment index consistently increased as oxygen tension fell (Fig. 1).

FIGURE 1. This curve represents the average effect of airway hypoxia on the level of capillary recruitment in the upper lung in a series of nine consecutive dogs. The curve swings upward at the same oxygen tension that systemic arterial hypoxemia begins to occur. (From Wagner and Latham, 1975.)

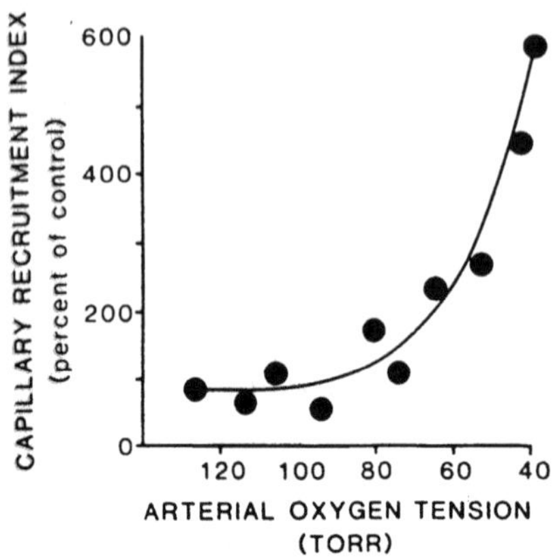

These observations lead to a series of studies designed to determine what caused the capillary recruitment. The obvious extrapulmonary causes, increased left atrial pressure and cardiac output, were eliminated because left atrial pressure was unchanged by hypoxia and cardiac output fell in some instances of hypoxia when there was substantial recruitment (Wagner and Latham, 1975; Wagner et al., 1979). In later work, we held output constant from control to hypoxia and found that recruitment occurred independently of total pulmonary blood flow (Capen and Wagner, 1982; Capen et al., 1981). Thus an intrapulmonary mechanism seemed to be the likely cause of the recruitment. Venoconstriction could certainly cause a retrograde rise in capillary pressure that would lead to recruitment. The well known elevation of pulmonary arterial pressure during hypoxia might also account for the recruitment by redistributing blood flow upward to our upper lung observation site. To differentiate between these potential causes, we directly measured pressure in both small pulmonary veins and arteries, made the animal hypoxic, and measured capillary recruitment. Then, while maintaining the hypoxic challenge at a constant level, a vasodilator, prostaglandin E_1, was infused to relieve whatever vasoconstriction had occurred (Capen and Wagner, 1982). The vasodilator caused a large reduction in the number of perfused capillaries. We could not measure any increase in pulmonary venous pressure during hypoxia, nor detect any effect of the vasodilator on vein pressure. Pulmonary artery pressure, however, fell to near control levels during vasodilator infusion. The plots of pressure versus recruitment (Fig. 2) showed no correlation with vein pressure but an impressive correlation with artery pressure. Therefore capillary recruitment during hypoxia did not correlate with cardiac output, pulmonary venous, or left atrial pressure, but did correlate with pulmonary arterial pressure. If recruitment was caused by a rise in capillary pressure, which it certainly must be, then how could the pressure rise in the capillaries which are located downstream of an upstream arterial constriction? It seemed more plausible that constriction of an artery feeding a capillary bed would have to be associated with reduced flow and derecruitment.

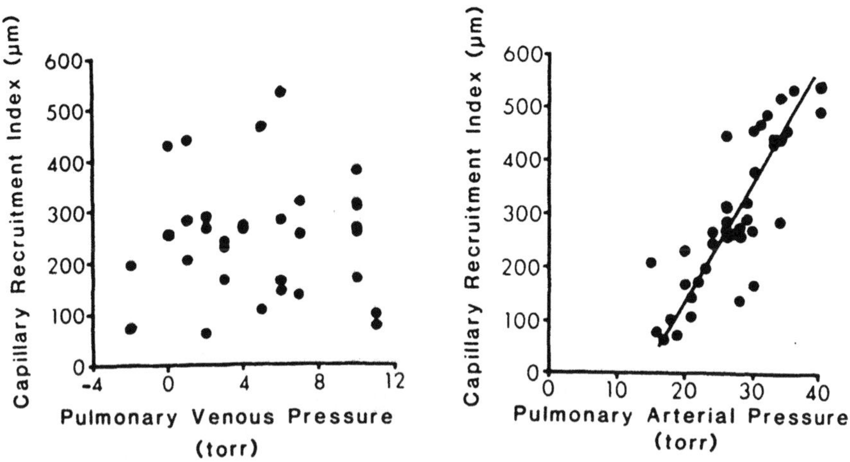

FIGURE 2. The correlation between capillary recruitment index and pulmonary venous pressure was not significant ($p = .9$), but was highly significantly correlated with pulmonary arterial pressure ($p < .001$) for this group of 10 dogs. (From Wagner et al., 1979.)

To help visualize the pressure, flow, and resistance relationships in the pulmonary circulation we developed a model. In some ways the model is undeniably too simple, but it is compatible with our data and has suggested a number of further experiments. The model is based on an apartment building and what happens when the tenants all try to take a shower at 7:00 AM (Fig. 3, left). When everyone opens their shower control valves simultaneously, the dwellers on the lower floors benefit from a rapid flow of water through a fully recruited shower head, analogous to the capillary bed in the lower lung. The upper floor tenants are confounded by a derecruited shower head. To this point, the model reproduces in part the elegant pulmonary circulatory hydrostatic zone model developed by Permutt et al. (1965) and West (1965) to describe the hydrodynamic distribution of blood flow under normal conditions. To extend the apartment house model, let us suppose that there is a meeting in which all tenants agree to open their control valves only partially (Fig.3, right). This condition represents generalized pulmonary arterial vasoconstriction. With higher pressure available, water is redistributed to the upper floors and holes in the shower heads are recruited. The paradox is solved of how capillaries can be recruited downstream from an upstream constriction, because some water is available to flow past the constriction in the control valves to the upper halves (Fig. 3, right), whereas no water had been available to flow past the wide open control valve (Fig. 3, left) under control conditions.

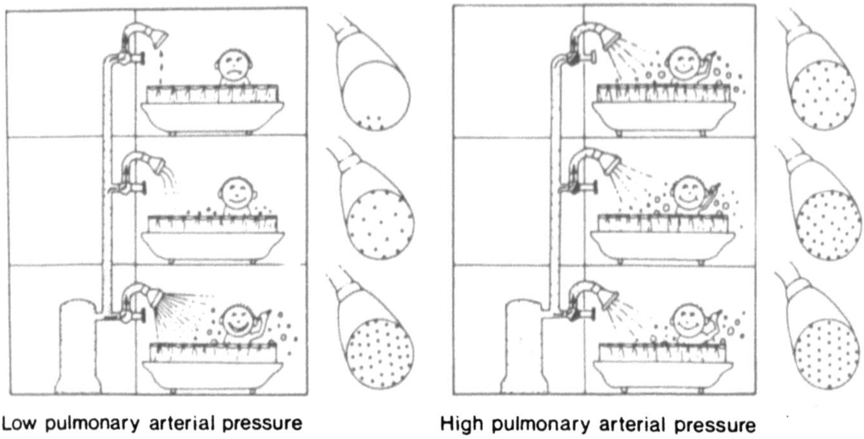

FIGURE 3. In this model the apartment house represents the lung and the shower heads the capillary bed. Capillary recruitment is analogous to recruitment of the holes in the shower head. Under normal conditions, the shower heads are not fully recruited in the upper stories and flow is brisk in the lower stories (*left panel*). During hypoxia, arterial vasoconstriction (represented by the partially closed control valves in the *right panel*) causes pressure to be elevated and flow to be redistributed to the upper shower heads, where recruitment occurs.

The model predicts, with conditions of steady total flow, analogous to lack of the cardiac output changes during hypoxia, and an evenly distributed constriction of all control valves, that there will be upward redistribution of flow which could result in an overall gain in the total number of holes in the shower heads via recruitment. Under these conditions, the extra flow to the upper shower heads must come from the lower apartments. The lesser flow in the lower apartments, however, need not lead to local derecruitment; rather the extra water could come from a reduction in the velocity of the water passing through the lower shower heads, analogous to slower capillary transit times. If this reasoning is correct, then the model predicts that in the lung there should be an increase in total capillary volume during the increased pulmonary arterial pressure associated with airway hypoxia.

To test this prediction, it was necessary to determine the effect of hypoxia on total pulmonary capillary volume. To do this, we measured the diffusing capacity of the lung for carbon monoxide (Capen et al., 1981). A diffusing capacity increase during hypoxia, however, could reflect either increased capillary volume (recruitment), or less competition from oxygen for hemoglobin binding sites, which would alter the reaction rate between carbon monoxide and hemoglobin (θ), or some combination of the two. To determine the effect of recruitment alone, the vasodilator prostaglandin E_1 was infused while airway hypoxia was held at a constant level. The resultant decrease in pulmonary artery pressure caused, as expected from earlier work, capillary derecruitment (Fig. 4, left). Diffusing capacity also decreased (Fig. 4, right). By assuming no change in membrane diffusing capacity, and by having held θ constant by keeping the level of hypoxia constant, the experiment showed that there was a net gain in capillary volume, most likely through capillary recruitment. Such an increase in gas exchange surface area would be advantageous during whole lung hypoxia.

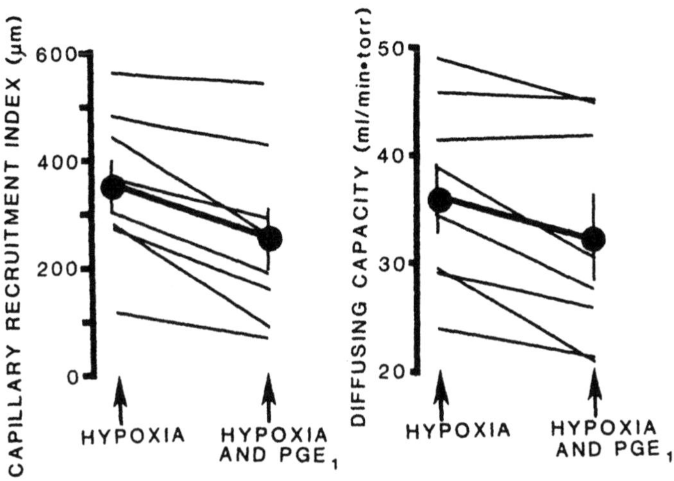

FIGURE 4. Capillary recruitment index and the diffusing capacity of the lung for carbon monoxide are plotted for a group of eight dogs. The thin lines are data from individual dogs. Heavy lines are group means and standard errors. Under these conditions hypoxia was held constant and recruitment was varied by infusing the vasodilator prostaglandin E_1. The derecruitment resulting from the vasodilator was associated with a fall in diffusing capacity. (From Capen, 1979.)

The apartment house model predicts upward redistribution of flow with constriction and downward redistribution with dilation. There is evidence that hypoxia causes upward redistribution of pulmonary blood flow both acutely in anesthetized dogs (Dugard and Naimark, 1967) and in man native to high altitude (Dawson and Grover, 1974). To determine what vasodilation did to blood flow distribution in our preparation, we injected radiolabelled 15 μm microspheres into the right atrium and measured the location of the wedged spheres in the lungs during hypoxia and hypoxia plus prostaglandin E_1. As expected, hypoxia caused microsphere distribution to be relatively even from top to bottom (Fig. 5). The vasodilator caused the curve to rotate (Fig. 5) indicating that blood flow diminished in the upper lung and returned to high levels in the lower lung. This seems convincing evidence that upward redistribution of blood flow existed in our preparations and is a likely explanation for capillary recruitment in the upper lung during hypoxia.

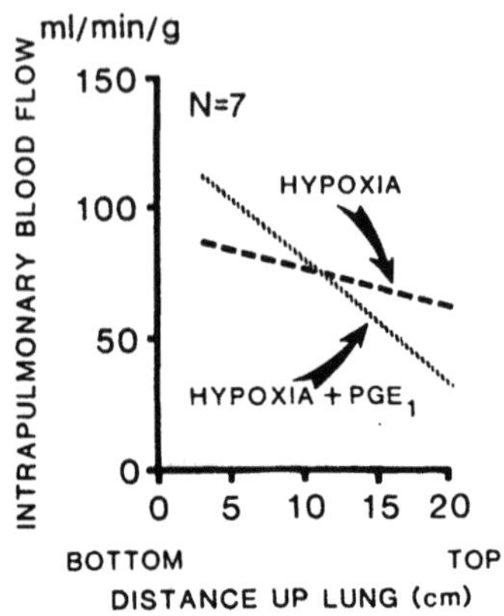

FIGURE 5. The intrapulmonary distribution of blood flow determined by radioactive microsphere distribution per gram of dry lung tissue plotted against distance up from the bottom of the lung ($n = 7$). During hypoxia the blood flow is fairly even from bottom to top, but when the vasodilator prostaglandin E_1 is infused to relieve the pulmonary arterial constriction the curve rotates, causing the slope to be steeper, an indication that blood flow is redistributed toward the bottom of the lung. The slopes of the lines are different from each other ($p < .05$). (From Capen and Wagner, 1982.)

From these data, the response of the pulmonary circulation can be summarized in the following way: hypoxia --> pulmonary arterial constriction --> increased pulmonary arterial pressure --> upward redistribution of blood flow --> capillary recruitment --> increased surface area for gas exchange. It is not certain how much improvement in arterial oxygen tension might occur from the increase in gas exchange surface area provided by the recruited capillaries. As best we can calculate it might not exceed 5 Torr in a normal lung, although it could be substantially more in a heterogeneously diseased lung. In any case, whenever arterial oxygen tension is below 30 or 40 torr, any addition would be welcome.

One of the main conclusions that we draw from this series of studies has to do with control of the pulmonary circulation. This and other work suggests that capillary recruitment is a passive event resulting from increased pressure in the capillary bed. Presumably the pressure rise can be from any source. For example, a downstream increase in resistance causing a retrograde rise in capillary pressure would be expected to cause recruitment just as readily as an upward redistribution of blood flow causes recruitment in the case of whole lung hypoxia. Since, however, we found no evidence that venoconstriction played an active role in capillary recruitment, we believe that changes in pulmonary arterial pressure play the dominant role in controlling whole lung hemodynamics.

References

Capen RL, Latham LP, and WW Wagner Jr. (1981) Diffusing capacity of the lung during hypoxia: the role of capillary recruitment. *J Appl Physiol* 50:165-171.

Capen RL, Wagner WW Jr. (1982) Intrapulmonary blood flow redistribution during hypoxia increases gas exchange surface area. *J Appl Physiol* 52:1575-1581.

Dugard A, Naimark, A. (1967) Effect of hypoxia on distribution of pulmonary blood flow. *J Appl Physiol* 23:663-671.

Dawson A, Grover RF. Regional lung function in natives and long-term residents at 3,100 m altitude. *J Appl Physiol* 36:294-298.

Permutt S, Bromberger-Barnea B, Bane HN. (1965) Alveolar pressure, pulmonary venous pressure, and the vascular waterfall. *Med Thorac* 22:118-131.

Wagner WW Jr, Filley GF. (1965) Microscopic observation of the lung in vivo. *Vasc Dis* 2:229-241.

Wagner WW Jr. (1969) Pulmonary microcirculatory observations in vivo under physiological conditions. *J Appl Physiol* 26:375-377.

Wagner WW Jr, Latham LP. (1975) Airway hypoxia causes pulmonary capillary recruitment in the dog. *J Appl Physiol* 39:900-905.

Wagner WW Jr, Latham LP, Gillespie MN, Guenther J, Capen RL. (1982) Red cell transit times across pulmonary capillaries. *Science* 218:379-380.

Wagner WW Jr, Latham LP, Capen RL. (1979) Capillary recruitment during airway hypoxia: the role of pulmonary artery pressure. *J Appl Physiol* 47:383-387.

West JB (1965) *Ventilation/blood flow and Gas Exchange*. Blackwell, Oxford.

Pulsatile pulmonary capillary pressure measured with the arterial occlusion technique

Jean-Michel Maarek and H.K. Chang

Department of Biomedical Engineering, University of Southern California, Los Angeles CA 90080–1451

1 Introduction

Pulsatile flow waves have been known to exist in the pulmonary capillary bed since the introduction of the N_2O bolus technique [5, 10]. Although direct measurement of pressure pulsatility in the lung microvessels has yet to be performed, indirect methods have been used to assess the cyclic variations of the pulmonary capillary pressure during pulsatile perfusion [9, 7]. Recently, it has been suggested [3, 13, 6] that these periodic oscillations could be traced by measuring the vascular occlusion pressures [4, 2, 11, 1] repeatedly during the pulsatile pressure cycle. We describe an application of this technique in isolated left lower lobes (LLL) of canine lungs perfused by means of a pulsatile blood pump.

2 Experiments and data analysis

The experimental measurements were carried out on seven anesthetized and mechanically ventilated mongrel dogs. After a left thoracotomy, the lower lobe of the left lung was isolated *in-situ*. The lobar artery and vein were cannulated [4, 6] to allow for the extracorporeal perfusion of the LLL. A pulsatile blood pump was used to circulate autologous blood between the LLL and a heated reservoir raised to the level of the top of the lobe. We measured the pressures in the lobar artery (Pa) and in the lobar vein (Pv) with identical transducers referenced to the top of the lung. An ultrasonic flow probe at the venous outflow of the LLL was used to evaluate the mean flow rate through the lobar vasculature.

After the experimental preparation had stabilized, the pulsatile flow pump was set at one of four frequencies: 36, 54, 72, or 90 beats/min.

The pump stroke was adjusted accordingly to keep the mean perfusion rate at 0.37 l/min. Arterial occlusions (AO) were then performed at 16 consecutive instants which completely covered the cycle of the perfusion pump. Double occlusion (DO) pressures were also measured at one out of two instants of AO. The sequence of measurements was then repeated for the three remaining frequencies. To investigate the interactions between capillary pulsatility and vasoconstriction, serotonin was infused in 6 lobes. The dose of serotonin (190 $\pm$ 12 μg/min) approximately doubled the lobar vascular resistance. The measurements of the vascular occlusion pressures were repeated at 36 and 72 beats/min using the same instants of occlusion as during the baseline measurements. In 4 lobes, we also increased the venous pressure from 1 to 10 mm hg and repeated the sequences of occlusion measurements at 36 and 72 beats/min.

For each occlusion maneuver, the pressures signals (Pa and Pv) and the flow signal (Q) were sampled on a computer during 15 seconds which covered several pump cycles before the occlusion and approximately 10 seconds of occlusion data. The arterial occlusion pressure (Pa0) was estimated by back-extrapolating to the instant of occlusion a single exponential fit of a segment of the AO curve extending from 0.3 s to 2.3 s after the occlusion [3, 6]. Double occlusion pressure (Pdo) was computed as the common value of pressures Pa and Pv in the period between the first and fourth seconds which followed a DO maneuver.

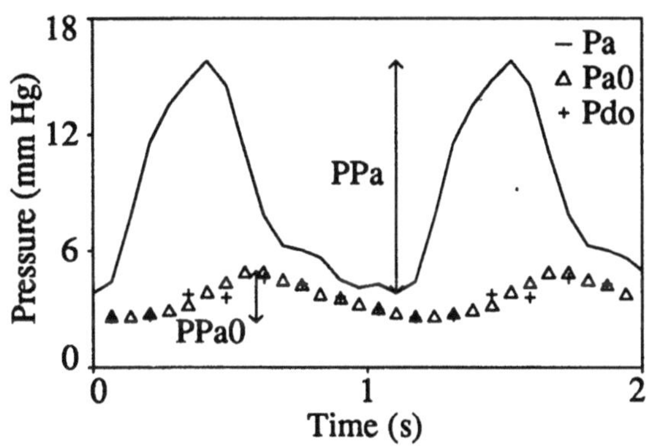

Figure 1: Capillary pressure wave reconstructed from multiple AO and DO pressure measurements.

Figure 1 presents plots of pressures Pa0 and Pdo relocated in the pump cycle with respect to the corresponding instants of occlusion. Smooth oscillatory patterns lagging behind the pulmonary arterial pressure contour resulted from this reconstruction. The peak-to-peak variations of the occlusion pressure waves were designated as pulse pressures PPa0 and PPdo.

The pulsatility transmission ratio (TR) was defined as the ratio of PPa0 over the pulse pressure in the pulmonary artery (PPa). Furthermore, for each of the pump frequencies, we computed the Fourier transforms of the pulmonary arterial pressure wave and of arterial occlusion pressure wave. These quantities were used to derive the transfer function of the arterial occlusion pressure at the four studied frequencies [8]. Such a function describes the amplitude (modulus) and phase relationships between the arterial pressure wave considered as the input to a linear system and the arterial occlusion pressure wave considered as the corresponding output.

3 Results

Pulse pressures PPa0 and PPdo were on the average equal which indicated that the AO and DO techniques yielded similar estimates of the pressure pulsatility in the pulmonary microvessels. The arterial occlusion pulse pressure (PPa0) decreased regularly from 3.6 (± 0.4) mm Hg (mean±SEM) to 1.7 (± 0.3) mm Hg as the pump frequency (f) was raised from 36 to 90 beats/min (figure 2).

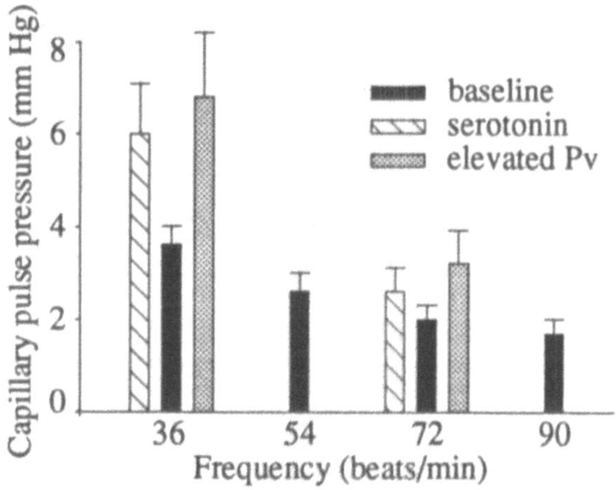

Figure 2: Pulse pressure (PPa0) in the pulmonary capillaries estimated from arterial occlusion pressure wave. Under baseline conditions, PPa0 decreased as the frequency was increased. Infusion of serotonin and elevation of Pv increased the values of PPa0 measured at 36 and 72 beats/min.

The arterial pulse pressure (PPa) decreased moderately between 15.5 (± 2.7) and 12.8 (± 1.7) mm Hg as f was increased and the pump stroke was reduced. Overall, the transmission ratio TR significantly decreased from 0.25 (± 0.02) to 0.14 (± 0.02) as f was increased (figure 3). The modulus of the transfer function at the site of Pa0 decreased from 0.31 (± 0.03) to 0.18

(± 0.03) as the frequency was increased from 36 to 90 beats/min (figure 4, left). Its variations were similar to those of the transmission ratio TR.

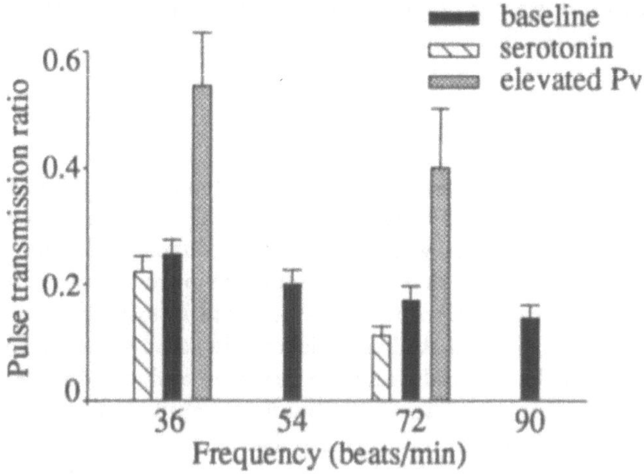

Figure 3: Transmission of pulsatility to the microvasculature decreased when the frequency of the arterial pressure waveform increased. Pulse transmission ratio TR was nearly unchanged during the infusion of serotonin; it increased when the venous pressure was raised to 10 mm Hg.

The phase (figure 4, right) decreased from -47 (± 3) ° to -82 (± 3) ° as the pump frequency was raised. A negative phase for all frequencies indicated that the Pa0 waveform was always delayed with respect to the pulmonary artery pressure wave.

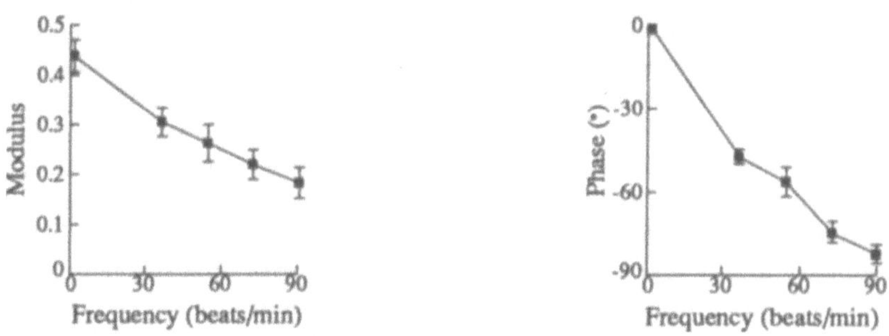

Figure 4: Modulus (left) and phase (right) of the pressure transfer function at the site of arterial occlusion. Transfer function indicated a larger attenuation and a larger phase lag at the higher frequencies.

Serotonin infusion and elevation of Pv to 10 mm Hg both resulted in an increase of pulse pressure PPa0 as compared to its baseline values (figure

2). Arterial pulse pressure PPa increased following the infusion of serotonin whereas it decreased when Pv was raised. As a result, the pressure transmission ratio TR was approximately unchanged in the former condition whereas it increased in the latter (figure 3).

4 Discussion

In earlier experiments performed in isolated lung lobes [6] and in semi-intact lungs [3], we showed that the cycle-averaged arterial occlusion and double occlusion pressures were nearly equal during pulsatile perfusion. Our present results show that throughout the pulsatile pump cycle, the arterial and double occlusion pressures oscillate in synchrony and remain close to each other. Several studies performed in steady-flow-perfused lungs have shown that the arterial occlusion pressure and the double occlusion pressure were nearly equal to the pulmonary capillary filtration pressure [11, 1]. Their phasic variations during pulsatile perfusion would be indicative of the presence of pulsatile pressure waves in the pulmonary capillaries.

Yamada et al. [13] recently performed electrocardiogram-controlled arterial occlusions in intact lung lobes perfused by the heart. These authors reported phasic variations of the estimated pulmonary capillary pressure lagging behind the pulmonary arterial wave. When the average heart rate was 118 beats/min, the peak-to-peak amplitude of the AO pressure waves averaged 2.0 mm Hg which corresponded to a transmission ratio of 0.26. The results obtained in our LLL preparation suggest a smaller magnitude of the microvascular pulse pressure at that frequency since we found that PPa0 and TR measured at 90 beats/min were around 1.7 mm Hg and 0.14, respectively. Differences in the levels of the venous pressure and in the experimental preparations probably account for the quantitative differences between our results and those of Yamada and colleagues. Qualitatively however, the two sets of measurements are in agreement. They support the existence of noticeable pulmonary capillary pressure waves during pulsatile flow perfusion. Pressure pulsatility in lung microvasculature was also reported by Maloney et al. [7]. In an isolated lung preparation in zone 2 conditions, these authors estimated that sinusoidal pressure waves in the pulmonary artery reached the collapsible vessels with 35% of their amplitude at 36 beats/min and with 20% of their amplitude at 90 beats/min. At these frequencies, we obtained comparable levels of attenuation from the modulus of the transfer function of the arterial occlusion pressure.

Increasing the frequency of the arterial pressure waveform resulted in a smaller amplitude for the capillary pressure waves. The transmission of pressure waves in the vasculature is a function of the complex wave velocity which describes the changes in amplitude and phase of a travelling wave [8]. The complex wave velocity is frequency-dependent which means that pressure waves at different frequencies are attenuated and delayed by different amounts. Reflected waves from the periphery also contribute to the dispersion and attenuation of the incident pressure waves [12]. Using a linear distributed model of the pulmonary vasculature, Wiener et al. [12] demonstrated that the harmonics of the arterial pressure wave were attenuated in different generations of pulmonary vessels. All the pressure harmonics appeared considerably attenuated in the microvessels. In these conditions, the capillary pressure wave would be rounder and smaller in magnitude than the pulmonary arterial waveform.

We observed that vasoconstriction induced by serotonin resulted in markedly increased oscillations for the vascular occlusion pressure contours. This probably reflected a large increase of the pulmonary arterial pulse pressure since the pressure transmission ratio did not change significantly during the infusion of serotonin. Previous experiments have shown that the transmission of flow waves to the pulmonary capillary bed was not altered when serotonin was infused [10]. Thus, it would seem that vasoconstriction induced by serotonin does not modify the transmission properties of the pulmonary vasculature. In contrast, the pressure transmission ratio appeared larger after elevation of the venous pressure to 10 mm Hg. The pulmonary vasculature was probably dilated and stiffer as a result of the increased distending pressure [8]. In such conditions, the transmission of pulsatile pressure waves would have occurred with less attenuation.

In summary, we have shown that the pulmonary capillary pressure oscillated during pulsatile flow perfusion and that these oscillations could be reconstructed through repeated measurements of the arterial or the double occlusion pressure. The capillary pressure waveform appeared delayed with respect to the pulmonary arterial wave and smoothened by the dampening properties of the pulmonary vascular bed. The presence of pulsatile waves in the pulmonary capillaries could have physiological consequences on the rate of fluid filtration across capillary wall and on the dynamics of gas exchange. Although the present experiments do not enable us to speculate on these issues, they provide us with techniques which could be used for these investigations.

Acknowledgement: This work was supported by NHLBI Grant HL 36908.

References

[1] Bshouty, Z., J. Ali, and M. Younes. Arterial occlusion versus isofiltration capillary pressures during very high flow. *J. Appl. Physiol.* 62: 1174–1178, 1987.

[2] Dawson, C.A., J.H. Linehan, and D.A. Rickaby. Pulmonary microcirculatory hemodynamics. *Ann. N.Y. Acad. Sci.* 384: 90–106, 1982.

[3] Hakim, T.S., J.M. Maarek, and H.K. Chang. Estimation of pulmonary capillary pressure in intact dog lungs using the arterial occlusion technique. *Am. Rev. Resp. Dis.* 140: 217–224, 1989.

[4] Hakim, T.S., R.P. Michel, and H.K. Chang. Partitioning of pulmonary vascular resistance in dogs by arterial and venous occlusion. *J. Appl. Physiol.* 52: 710–715, 1982.

[5] Lee, G. de J., and A.B. DuBois. Pulmonary capillary blood flow in man. *J. Clin. Invest.* 34: 1380–1390, 1955.

[6] Maarek, J.M., T.S. Hakim, and H.K. Chang. Analysis of pulmonary arterial pressure profile after occlusion of pulsatile blood flow. *J. Appl. Physiol.* 68: 761–769, 1990.

[7] Maloney, J.E., D.H. Bergel, J.B. Glazier, J.M.B. Hughes and J.B. West. Transmission of pulsatile blood pressure and flow through the isolated lung. *Circ. Res.* 23: 11–23, 1968.

[8] Milnor, W.R. Hemodynamics. Williams & Wilkins. Baltimore, 1982.

[9] Morkin, E., J.A. Collins, H.S. Goldman, and A.P. Fishman. Pattern of blood flow in the pulmonary veins of the dog. *J. Appl. Physiol.* 20: 1118–1128, 1965.

[10] Reuben, S.R., J.P. Swadling, B.J. Gersh, and G. de J. Lee. Impedance and transmission properties of the pulmonary arterial system. *Cardiovasc. Res.* 5: 1–9, 1971.

[11] Rippe, B., J.C. Parker, M.I. Townsley, N.A. Mortillaro, and A.E. Taylor. Segmental vascular resistances and compliances in dog lung. *J. Appl. Physiol.* 62: 1206–1215, 1987.

[12] Wiener F., E. Morkin, R. Skalak, and A.P. Fishman. Wave propagation in the pulmonary circulation. *Circ. Res.* 19: 834–850, 1966.

[13] Yamada, Y., M. Suzukawa, M. Chinzei, T. Chinzei, N. Kawahara, K. Suwa, and K. Numata. Phasic capillary pressure determined by arterial occlusion in intact dog lung lobes. *J. Appl. Physiol.* 67: 2205–2211, 1989.

Sites of Pulmonary Vasoconstriction: Indirect and Direct Measurements.

John H. Linehan and Christopher A. Dawson, Marquette University, Milwaukee, WI 53233; Medical College of Wisconsin, Milwaukee, WI 53226 and Zablocki VA Medical Center, Milwaukee, WI 53295

Because pulmonary capillary pressure is an important factor in the fluid balance of the lungs, there has been considerable interest in methods for determining it and the arteriovenous sites of pulmonary vasoconstriction. Since pulmonary arterial and venous pressures but not capillary pressure can be measured directly, the methods have, in general, been indirect (Dawson, 1984; Dawson et al., 1989). In our laboratory, we have used two complementary indirect methods; the vascular occlusion method (Linehan et al., 1982) and the low viscosity bolus method (Dawson et al., 1988).

The transient pressure and flow data following rapid occlusion of the venous outflow from an isolated perfused lung contain information about the arterial-to-venous distribution of vascular resistance (R) relative to the distribution of vascular compliance (C). Following rapid occlusion of the vein, the venous pressure jumps quickly to a pressure somewhere between the preocclusion arterial and venous pressures. Subsequently, the vascular bed fills with blood. As a result, the arterial and venous pressures begin to rise more slowly. Examples of the pressure versus time curves from venous occlusion are shown in the upper panel of figure 1 (Dawson, 1984 and Dawson et al., 1982). In this figure, the left lower lobe of the dog lung is perfused at constant flow during control and when either serotonin or histamine is infused. The magnitude of the rapid jump in venous pressure in relation to the total arterial – venous pressure difference varies depending on the distribution of the vascular resistance relative to vascular compliance. A simple hemodynamic model that can explain the most obvious features of the pressure curves is

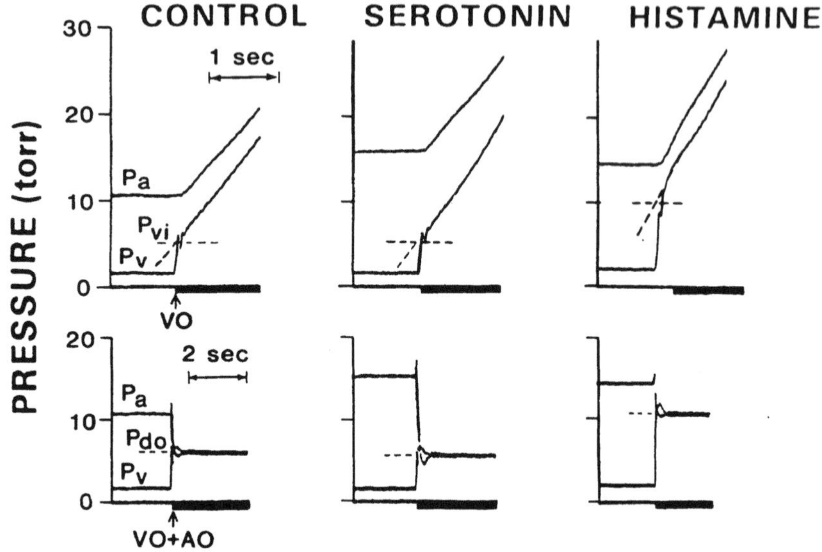

Figure 1. The arterial P_a and venous P_v pressure responses following venous occlusion or simultaneous venous and arterial occlusion in an isolated dog lung lobe. The horizontal dashed line designated P_{vi} is the pressure obtained by extrapolating the P_v curve back to the instant of venous occlusion. The horizontal dashed line designated P_{do} is the equilibrium pressure following simultaneous arterial and venous occlusion. The relationships between P_a, P_v, P_{vi}, and P_{do} are altered by vasoconstriction induced by serotonin and histamine infusion. (From Dawson, 1984 and Dawson et al., 1982).

shown as an electrical analog T section in figure 2. In this model, when flow through the downstream resistance (R_v) stops as a result of opening the downstream switch (analog of venous occlusion), the P_v jumps to the preocclusion pressure P_{vi} at the central compliance, C_L. Since the flow through R_a remains constant the pressure P_{vi} of the volume storage element will instantly equal P_v and thereafter increase linearly with time. The pressure, P_a, upstream of R_a will rise in parallel with P_v according to $P_a = P_v + R_a\dot{Q}$. Thus, we interpreted a rise in P_{vi} as reflecting an increase in venous resistance while an increase in P_a without a concomitant change in P_{vi} is an increase in arterial resistance. Further, more-detailed interpretations of the venous occlusion data have been discussed in (Linehan, et al., 1982,

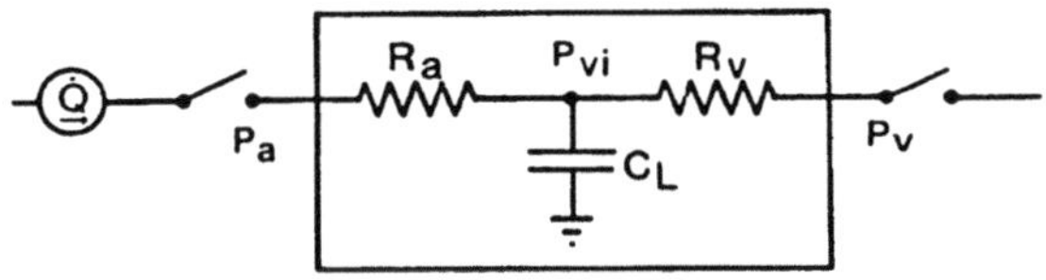

Figure 2. An electrical analogue T section used to explain the overall features of the arterial and venous pressure responses to occlusion.

1983, and 1988, Bronikowski, et al., 1984 and 1985. As shown in the bottom panel of figure 1, the simultaneous occlusion of both the artery and vein produces a rapid increase and a rapid decrease in P_V and P_a, respectively, to a common final pressure P_{do}. These data can also be simply interpreted by the model in figure 2. On balance, it can be seen that serotonin is primarily an arterial constrictor upstream from the locus of vascular compliance while histamine is a venous constrictor downstream from the locus of vascular compliance.

The low viscosity bolus method attempts to quantify the longitudinal distribution of vascular resistance with respect to cumulative blood volume. The essence of our development of this method has been described previously (Dawson et al., 1988). The basic approach involves the measurement of the change in the arteriovenous pressure gradient following the introduction of a bolus of saline, plasma, or blood diluted with saline or plasma into the blood flowing into the pulmonary artery of a pump-perfused lung. The pump provides a constant flow so that time-varying changes in the a-v pressure difference reflect the decrease in the pulmonary vascular resistance caused by the bolus. The magnitude of the instantaneous decrease in the a-v pressure difference with respect to the preinjection value follows a time course that depends on the longitudinal location of the bolus within the vascular bed at a given time and the size of the preinjection vascular resistance at that location. Thus, the decrease in the a-v pressure difference will tend to be largest when the bolus is located in the regions of highest (preinjection) resistance. Figure 3 shows how the shape of the pressure curve can change as a result of vasoconstriction. The objective, then, is to determine the longitudinal distribution of vascular resistance with respect to vascular volume from the arterial inlet to venous outlet from these pressure difference versus time curves that result from the convective passage of the bolus.

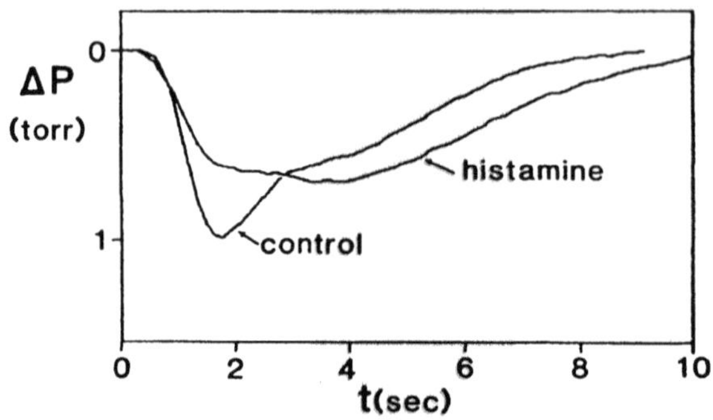

Figure 3. The change in the arteriovenous pressure difference, $\Delta(P_a-P_v)$ referred to as ΔP on this graph, following the introduction of a 2.5-ml bolus of saline-diluted blood through an isolated dog lung lobe perfused with blood at a flow rate of about 6 ml/sec. The infusion of histamine caused a marked change in the shape of the curve due to a shift in the site of major resistance toward the venous end of the vascular bed. The hematocrits of the two boluses were adjusted so that even though the total vascular resistance was higher during histamine infusion the areas above the ΔP curves were similar.

The details of the mathematical model and methods of numerical analysis have been delineated elsewhere (Dawson et al., 1988). In principle the longitudinal distribution of vascular resistance is defined as a finite sum of individual serial resistances. The individual resistances are assumed to be the product of time-varying perfusate viscosities and time-invariant constants which themselves contain local geometrical equivalents (in analogy with Poiseuille's law). Specific knowledge of the geometry and local vascular resistance in the bed is not necessary as long as the resistance is linearly proportional to viscosity and the geometric constants do not vary with time. Although the number of equal serial volume segments is arbitrary, we have chosen 20. The results obtainable from the method are presented for histamine in figure 4. The staircase graphs representing the local resistance per unit volume versus cumulative vascular volume manifest the vasoconstrictor effects of histamine taken in relationship to control. As can be seen, the vascular resistance upstream from the mid-point of vascular volume is unaffected by histamine but the resistance downstream from the mid-volume point is increased relative to control. This result complements the interpretations of the effect of histamine on the venous occlusion experiments.

In contradistinction, when serotonin is infused and the pulmonary artery pressure increases, the low viscosity bolus technique gives the results shown in figure 5. In this case, serotonin

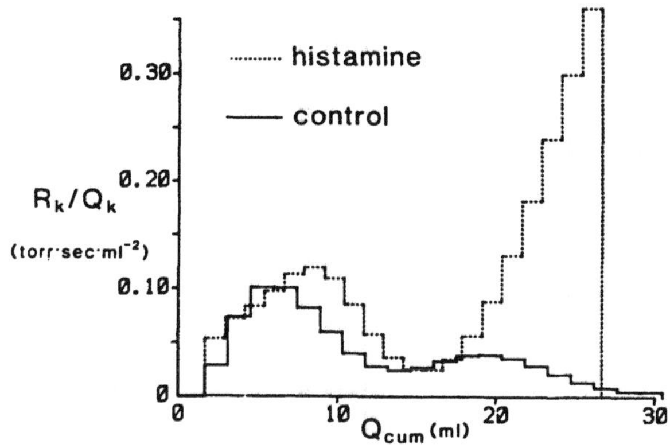

Figure 4. The resistance distribution obtained during histamine infusion compared with that obtained during normal control conditions. The vascular resistances is in terms of the segmental resistance per unit segmental volume. Q_{cum} is the cumulative vascular volume starting with the arterial inlet and extending to the venous outlet. Histamine caused a change in the bimodal distribution from upstream to primarily downstream constricting, presumably mainly venous constriction.

dramatically increases the pulmonary vascular resistance at the arterial end of the vascular bed defined by vascular volume. It is noted that the resistance distal to the mid-volume point is the same as for control. In figure 6, results obtained when the lung was made hypoxic are shown. The vascular resistance per unit vascular volume increases on the arterial side of the graph. This result appears, in the main, consistent with the serotonin response. However, closer examination of figures 5 and 6 suggests that while both hypoxia and serotonin increased arterial resistance, serotonin may have decreased the arterial compliance as well. This is based on the observation that during serotonin infusion the vascular segments having the highest resistance were at smaller values of Q_{cum} than during either control or hypoxic conditions. Thus, it appears that this method can distinguish between arterial and venous sites of vasomotion and provide some additional details about the effects of vasoconstrictor stimuli on the pulmonary vascular mechanics.

In general, the results of the two methods are consistent. To directly measure the sites of vasoconstriction, we have used an x-ray fluoroscope. Using a 40 µm focal spot x-ray source and an

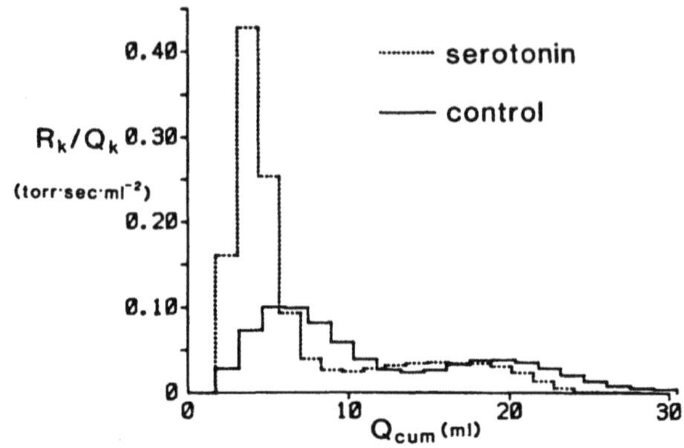

Figure 5. The resistance distribution obtained during serotonin infusion compared with that obtained during normal control conditions. Serotonin caused primarily upstream constriction, presumably arterial constriction. Same format as for Figure 4. (From Dawson et al., 1989, reprinted by courtesy of Marcel Dekker, Inc.)

x-ray camera, we record the passage of a bolus containing contrast media in a pump perfused lung under normal and vasoconstrictor conditions.

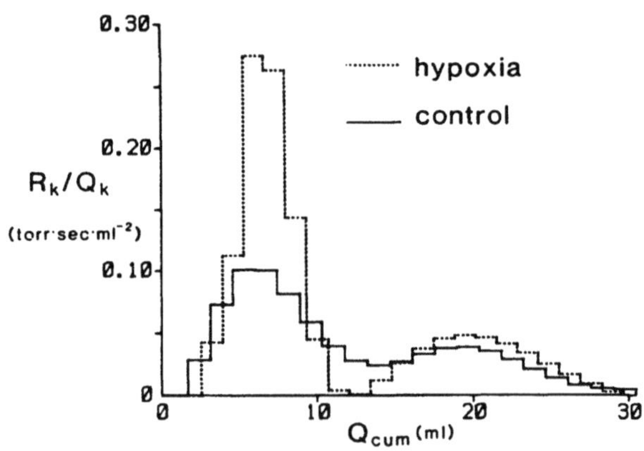

Figure 6. The resistance distribution obtained during hypoxia (P_AO_2 = 35 Torr) compared with that obtained during normal control conditions ((P_AO_2 = 111 Torr). Hypoxia caused primarily upstream constriction, presumably arterial constriction. Same format as for Figure 4.

As can be seen from figure 7, serotonin constricts all the arteries while hypoxia constricts only the smaller arteries. This difference is consistent with the results from the low viscosity bolus method, figures 5 and 6.

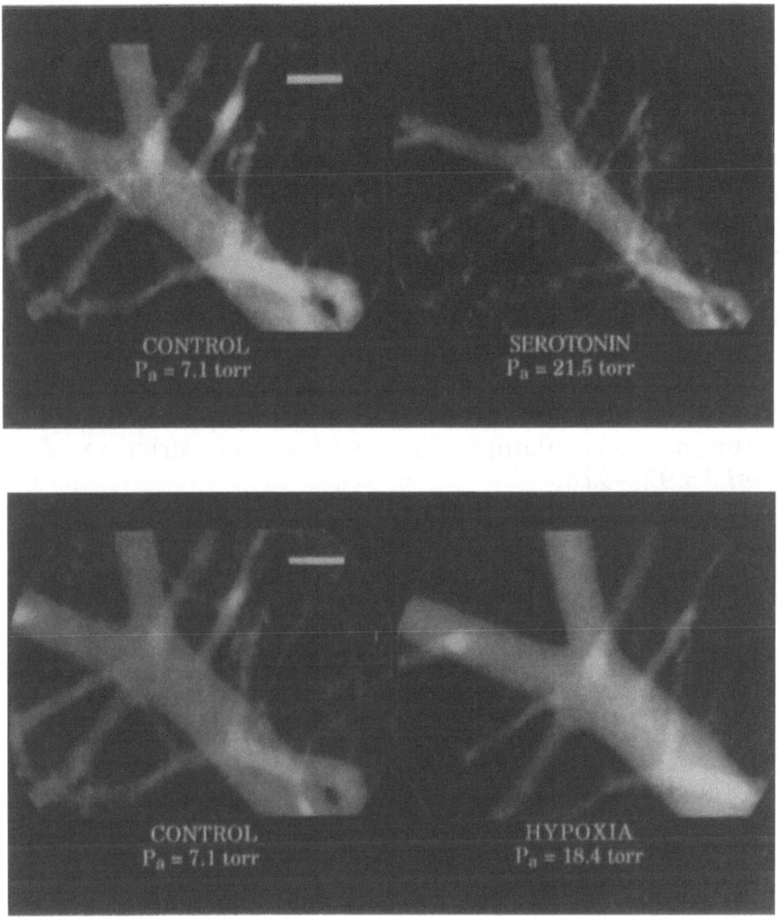

Figure 7. Fluoroscope images of the arteries of a pump perfused left lower dog lung lobe for control, serotonin infusion, and hypoxia. P_a is the lobar artery pressure.

Acknowledgment. This study was supported by National Heart, Lung and Blood Institute grant HL-19298 and the Research Service of the Veterans Administration.

References

Dawson CA (1984) Role of pulmonary vasomotion in physiology of the lung. *Physiol Rev* 64:544-616.

Dawson CA, Bronikowski TA, Linehan JH, Rickaby DA (1988) Distributions of vascular pressure and resistance in the lung. *J Appl Physiol* 64:274-284.

Dawson CA, Linehan JH, Bronikowski TA (1989) Pressure and flow in the pulmonary vascular bed. In EK Weir, JT Reeves (eds)*Pulmonary Vascular Physiology and Pathophysiology* Marcel Dekker, New York, pp 51-105.

Dawson CA, Linehan JH, Rickaby DA (1982) Pulmonary circulation hemodynamics. *Ann NY Acad Sci* 384:80-106.

Linehan JH, Dawson CA (1983) A three compartment model of the pulmonary vasculature: Effects of vasoconstriction. *J Appl Physiol* 55:923-928.

Linehan JH, Dawson CA, Rickaby DA (1982) Distribution of vascular resistance and compliance in a dog lung lobe. *J Appl Physiol* 53:158-168.

Linehan JH, deMora F, Bronikowski TA, Dawson CA (1988) Hemodynamic modeling of vascular occlusion experiments in cat lung. ASME.BED *Adv in Bioeng* 8:139-142.

THE USE OF MATHEMATICS AND ADVANCED TECHNOLOGY TO MEASURE AND EVALUATE LUNG FLUID EXCHANGE AND SOLUTE BALANCE

AN ERROR ANALYSIS OF PULMONARY VASCULAR PERMEABILITY MEASUREMENTS MADE WITH POSITRON EMISSION TOMOGRAPHY. D.P. Schuster, J. Markham, J. Kaplan, T. Warfel and M. Mintun. Washington University School of Medicine, St. Louis, MO.

Positron emission tomography (PET) is a powerful, quantitative, nuclear medicine imaging technique, useful for studying many problems in lung physiology and biochemistry (1). With PET, compounds are labeled with positron-emitting isotopes. After being administered either intravascularly or inhalationally, the tissue activity concentration of the isotope is determined with an imaging device similar in appearance to an X-ray computed tomography scanner. Multiple two-dimensional images are then reconstructed from the activity data and interpreted to represent a physiologic process of interest. PET derives its power from several factors: (1) the labeled compounds are themselves biologically important; (2) the isotope half-life is often sufficiently short that studies may be repeated if desirable; (3) the isotope tissue concentration can be determined quantitatively, accurately, and in many instances, noninvasively; and (4) the activity distribution can be located with great accuracy. Because of this latter feature in particular, the activity data can be presented regionally, in an image format, so that measurements may be correlated with other regionally specific measurements over time.

Radioactively-labeled compounds are used during PET to "trace" a physiologic or biochemical process. All isotopes used in PET decay by positron emission. After emission from the nucleus, the positron travels a few mm before it combines with an ambient electron. This interaction results in the annihilation of both original particles, with electromagnetic radiation given off. The annihilation radiation is released as two high-energy (511 keV) photons traveling in nearly opposite directions. These photons easily penetrate body tissue. Therefore, they can be recorded by suitable detectors placed on opposite sides of the subject. In modern PET devices, multiple radiation detectors are arranged about the subject and electronically linked to a computer capable of processing the enormous amounts of time-activity data produced during a typical PET study.

After all the activity data are collected, activity profiles are constructed from multiple projections taken at different angles. Each projection represents the regional tissue activity distribution of individual detector pairs. Combining these projections leads to a two-dimensional image of activity distribution within the field of view, each image representing a "slice" through the lung. The mathematics of this process are identical to that used in creating images by X-ray computed tomography. Multiple simultaneously obtained slices allow one to interpret the data three-dimensionally.

Pulmonary vascular permeability can be evaluated with PET by measuring the flux of a radiolabeled protein across the pulmonary endothelium(2-4). In this case, a positron-emitting, radiolabeled protein such as ^{68}Ga-transferrin is injected intravenously, and the activity measurements are made with sequential PET scans. The time required for the protein tracer to achieve equilibration between intravascular and extravascular spaces is dependent on both the local capillary endothelial permeability to the protein and the surface area available for efflux. Therefore, a surface area normalized transport rate constant (called the pulmonary transcapillary escape rate or PTCER), calculated from tissue-activity measurements obtained during a one hour period, is used as an index of pulmonary vascular permeability. Because PET cannot distinguish whether the activity from the radiolabeled protein originates from the intravascular or extravascular compartment, the accumulation of activity in the extravascular space is estimated by calculating the intravascular activity separately and subtracting it from the total measured value.

A simple two-compartmental model is used to evaluate the tissue and blood time-activity data (5,6). The equation that describes this model relates PET-measured tissue activity to activity measured in blood:

$$C_{PET} = C_{pl}\,(V_{pl}) + K_1 \int_0^T C_{pl} * e^{-k_2 t}\,dt$$

where C_{PET} is tissue activity measured by PET (counts/vol of lung), Cpl is plasma protein activity during the scan (counts/ml plasma), Vpl is plasma volume in any region of interest within the scan (ml), the asterisk is the convolution operator, and K_1 and k_2 are forward and reverse transport rate constants for radioactivity moving between vascular and extravascular compartments. Because the photons detected as a result of any positron emission are monoenergic, a second simultaneously administered vascular tracer cannot be used during PET. Instead, all the activity is assumed to be confined to the vascular pool at time zero (the time of intravenous administration), and vascular volume and rate constants between vascular and extravascular compartments are calculated simultaneously by iterative parameter estimation techniques (2).

In the lung, with varying amounts of lung tissue per unit volume, each rate constant would vary according to the surface area available for protein movement. However, assuming that regional blood or plasma volume will vary linearly with surface area, the rate constants can be normalized to changes in surface area by dividing each regional value by the regionally specific blood or plasma volume. Thus,

$$\text{PTCER} = K_1/\text{Vpl}$$

With this method of evaluating permeability, one must assume that protein movement across the endothelial barrier is passive, not active, and that pulmonary blood volume does not change during the course of the scan. Another assumption is that the regional estimate of blood (or plasma) volume is an appropriate index of surface area for normalizing the forward rate constant.

We have used computer simulation techniques to evaluate the validity of these assumptions. Tissue and blood time-activity curves were first generated to simulate the activity data that might be obtained by PET. These time-activity curves were then analyzed to produce estimates of PTCER. To produce the simulated data curves, values were assumed for the vascular volume within the lung tissue being analyzed, for the "true" PTCER, for the reverse rate constant (k_2), and for rate constants describing tracer clearance from blood. Specific values for these parameters represented average values from previous studies by us (2,3).

After generating these data curves, specific parameters used to produce them were altered in a systematic fashion to simulate potential errors in either the data collection or the analysis. All simulations were performed on a Concurrent Model 3205 Computer with program written in Fortran.

To examine the effect that errors in measuring the volume of the vascular compartment might have on the evaluation of vascular permeability, the tissue time-activity curve was first corrupted by adding activity assumed to be within blood vessels capable of exchanging this activity with the extravascular compartment but actually in blood vessels <u>in</u>capable of such exchange (e.g. all non-alveolar blood containing structures, such as large blood vessels or cardiac chambers).

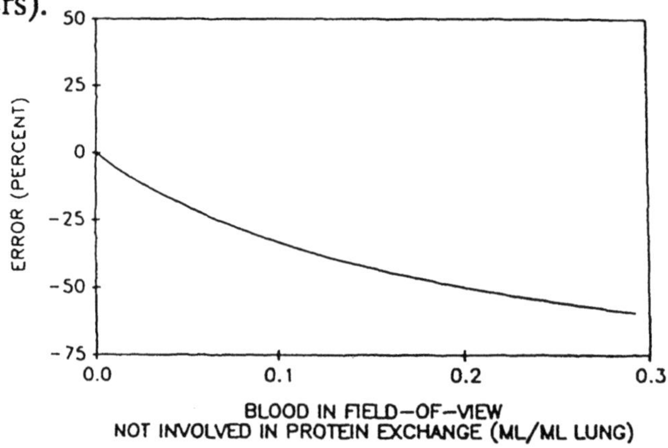

Figure 1. Errors in calculating PTCER if activity which is located in blood not involved in protein exchange is included. For this simulation, we assumed no error would occur if blood not involved in protein exchange could be completely excluded from the region-of-interest.

If blood not involved in vascular-extravascular compartmental protein exchange is included in the tissue activity measurement, then errors are to be expected in estimating PTCER. Including activity from non-exchanging blood (from, for example, non-capillary blood vessels) in the tissue-activity measurement causes PTCER to be underestimated by incorrectly estimating blood volume involved in tracer exchange, thereby increasing the denominator used in equations calculating PTCER. The magnitude of this effect is substantial, but depends on the ratio of real exchanging vascular volume to non-exchanging blood in the region of lung being analyzed.

Such errors should be minimized with PET because tomographic images are created which allow the investigator to place regions-of-interest away from the chest wall, larger vessels, and cardiac structures. These regions-of-interest, however, may still contain some blood in vessels not capable of exchanging tracer with the extravascular compartment. Thus, this type of error may be minimized but not entirely eliminated by the PET method.

In another analysis, the vascular volume of the lung region was assumed to change in a linear fashion over the actual data collection period. This simulation allowed us to evaluate the importance of assuming a constant vascular volume during data collection.

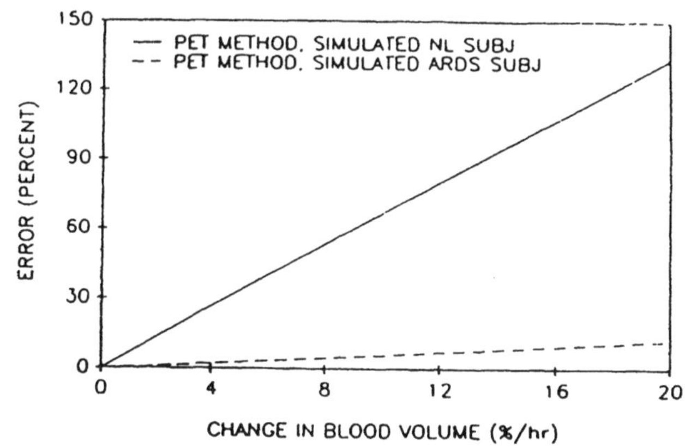

Figure 2. Errors in calculating PTCER as blood volume involved in exchange of label with the extravascular compartment <u>during the period of data collection</u> (one hour) is varied. The errors occur because the PET method assumes no such change in blood volume when calculating PTCER.

The results show that if blood volume changes occur <u>during the scan itself</u> and go undetected with PET method, significant errors in calculating PTCER will occur, particularly if PTCER is low (i.e. normal).

However, we don't believe this to be a major problem for three reasons. First, the absolute error for normal values of PTCER will be small compared with the percent error. Second, there is usually no a priori reason to believe that blood volume will change significantly during the scan period itself, especially in a stable or normal subject. Finally, such changes, should they occur, can be minimized by keeping the scan period relatively short. To simply calculate PTCER, the scan period can be limited to only 20 min (2). However, doing so requires that the data collection be started early after the administration of label, before k_2 affects the tissue protein time-activity measurement. In turn, this often requires that the blood time-activity curve be fitted to something other than a simple mono-exponential decay. We found empirically that a bi-exponential decay produces an acceptable fit to such data (2).

Finally, we evaluated the error which would occur if pulmonary blood flow is so low that equilibrium between the pulmonary artery and the regional tissue vascular compartment is not achieved before scanning begins.

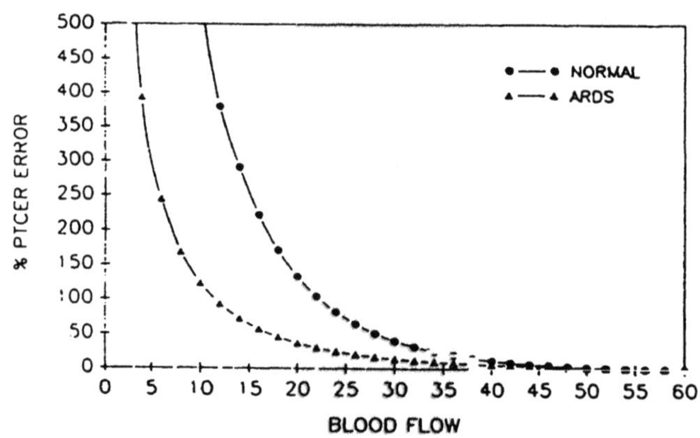

Figure 3. Errors in calculating PTCER as a function of regional pulmonary blood flow (ml/min/100 ml lung). At very low blood flows, the error is primarily due to mis-estimation of blood volume in the operational equation, as arterial blood activity (actually sampled) and regional pulmonary blood activity are no longer equivalent.

We found that significant errors can occur in calculating PTCER, particularly in tissue with normal permeability, leading to marked over-estimation of PTCER in such areas.

Thus, overall strategies to empirically minimize errors include carefully placing regions-of-interest, limiting scan duration, and including blood flow in the model equations if regional blood flow is less than 0.1 - 0.5 ml/min/ml lung.

REFERENCES

1. Schuster D.P. Positron emission tomography: theory, and its application to the study of lung disease. Am. Rev. Respir. Dis., 139:818-840, 1989.

2. Mintun M.A., D.R. Dennis, M.J. Welch, C.J. Mathias, and D.P. Schuster. Measurements of pulmonary vascular permeability with PET and gallium-68 transferrin. J. Nucl. Med. 28:1704-1716, 1987.

3. Calandrino F.S., D.J. Anderson, M.A. Mintun and D.P. Schuster. Pulmonary vascular permeability during the adult respiratory distress syndrome: a positron emission tomographic study. Am. Rev. Respir. Dis., 138:421-428, 1988.

4. Mintun M.S., T.E., Warfel, D.P. Schuster. Evaluating pulmonary vascular permeability with radiolabeled proteins: an error analysis. J. Appl. Physiol. In press.

5. Gorin A.B., J. Kohler, and G. DeNardo. Noninvasive measurement of pulmonary transvascular protein flux in normal man. J. Clin. Invest. 66:869-877, 1980.

6. Gorin, A.B., W.J. Weidner, R.H. Demling, and N.C. Staub. Noninvasive measurement of pulmonary transvascular protein flux in sheep. J. Appl. Physiol. 45:225-233, 1978.

EVALUATION OF LUNG VASCULAR PERMEABILITY BY EXTERNAL SCANNING OF GAMMA EMITTER ACTIVITY

Robert J. Roselli, Valerie J. Abernathy, William R. Riddle, Richard E. Parker, and N. Adrienne Pou, Departments of Biomedical Engineering, Radiological Sciences, and Medicine, Vanderbilt University, Nashville TN 37235.

Introduction

A minimally invasive method for measuring and analyzing lung transvascular protein flux would be useful for diagnosing and monitoring patients with lung injury. We have developed a method which detects movement of radiolabeled albumin from plasma to lung interstitium using sodium iodide detectors that are located outside the body, and positioned over the lung. The technique is similar to methods reported by Gorin (1,2), Prichard and Lee (3,4), and Dauber et al. (5).

Similar methods have been used to assess transvascular flux in humans (6,7), but in only one study has any attempt been made to verify the accuracy of the technique (1). Our objective was to test the external scan method using a highly controllable animal model in which direct measurements of transvascular protein flux could be measured.

Methods

A blood-perfused in situ sheep lung lymph preparation was used to test the accuracy of the external scan method (8). This preparation allows for comparison between externally detected transvascular protein movement and direct measurement of labeled protein in lung lymph. A catheter was surgically implanted into carotid artery and the caudal mediastinal lymph node was cannulated one week prior to the experiment. This was done to allow time for the lung to heal before beginning the experiment.

A schematic of the experimental setup is shown in Figure 1. One scintillation detector is positioned over the right lung and a second detector is positioned over a portion of the perfusion system. Cr-51 labeled red cells are injected into the perfusion system to permit corrections for blood volume changes within the monitored lung. The rate of accumulation of I-131 in the lung is monitored for at least a one-hour baseline period. This is followed by one of the following manipulations: a 15 cm H_2O elevation in microvascular pressure, a 15 cm H_2O elevation in alveolar pressure, infusion of air into the vascular system at a rate sufficient to double pulmonary vascular resistance, or infusion of perilla ketone at various doses.

A blood-corrected interstitial tracer mass to plasma tracer mass ratio (I/P) is computed from the radioactivity of I-131 and Cr-51 measured with the two detectors. A review of the analytical methods is given by Roselli and Riddle (9).

Results and Discussion

The radioactivity vs. time relationship during a control study is shown in Figure 2. The curve is generally linear, and the slope is related to albumin permeability of lung microvessels (9). We have found that this initial slope can be highly influenced by the presence of small amounts (less than 5%) of unbound radiolabel, and therefore all count rates are corrected for free tracer (10).

The I/P ratio correlates well with the measured I-131 albumin lymph to plasma concentration ratio. Two corrections must be made before a comparison can be made with the direct measurement. First, the mass I/P must be converted to a concentration I/P by multiplying by the ratio of plasma volume to interstitial volume available to albumin. Since protein is excluded from much of the lung interstitium (11) we used values between 1.3 and 2 for this ratio. Second, the interstitial concentration must be corrected for transport delay in lymph vessels.

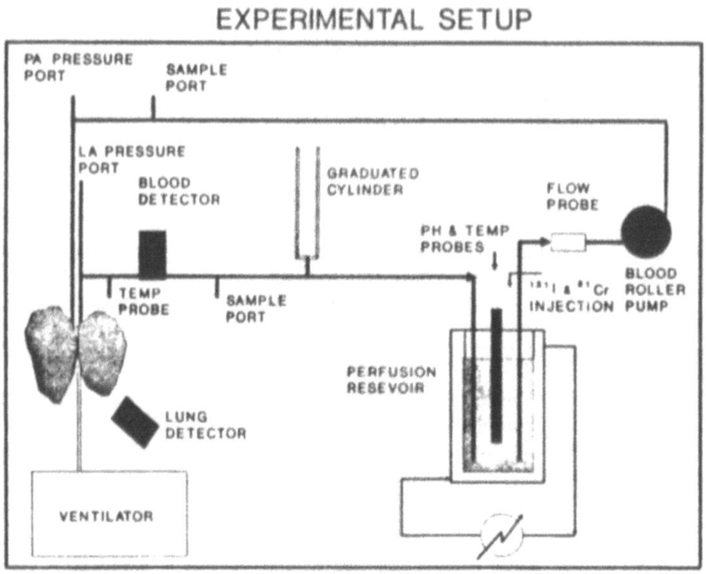

Figure 1. Schematic of apparatus used in external scan studies.

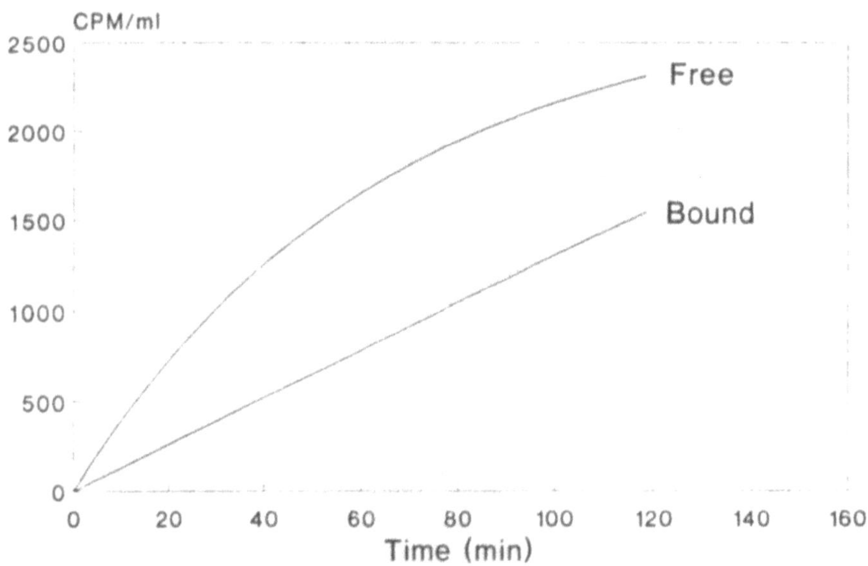

Figure 2. I-131 albumin detected externally during a control study.

This was done by assuming plug flow from initial lymphatics to the end of the lymph catheter, and selecting a value for the volume of these transit vessels. A delay time could then be calculated by dividing the lymphatic volume by the measured lymph flow. Best fit values for lymph volume was generally between 2-8 ml. A comparison between externally computed I/P and direct measurement of I/P, corrected as discussed above, is shown in Figure 3 for three studies.

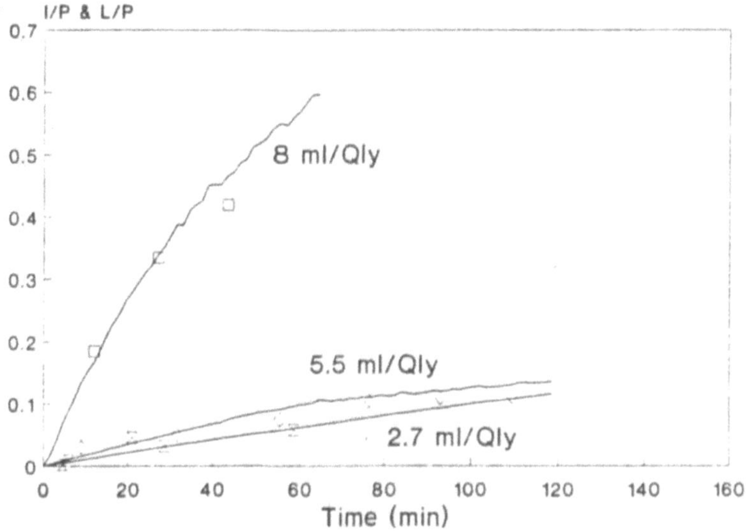

Figure 3. Comparison between measures of mass I/P computed from external scan (trace) and direct measurement in lymph (symbols). Lymph volume is 2 ml, and the ratio of plasma to interstitial volume is 2.0.

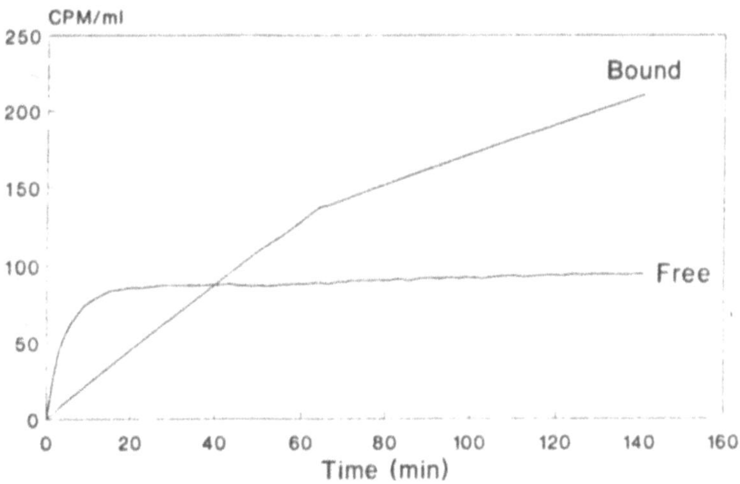

Figure 4. I-131 Albumin counts per minute per ml of lung volume (CPM/ml) following a 15 cm H_2O step change in microvascular pressure.

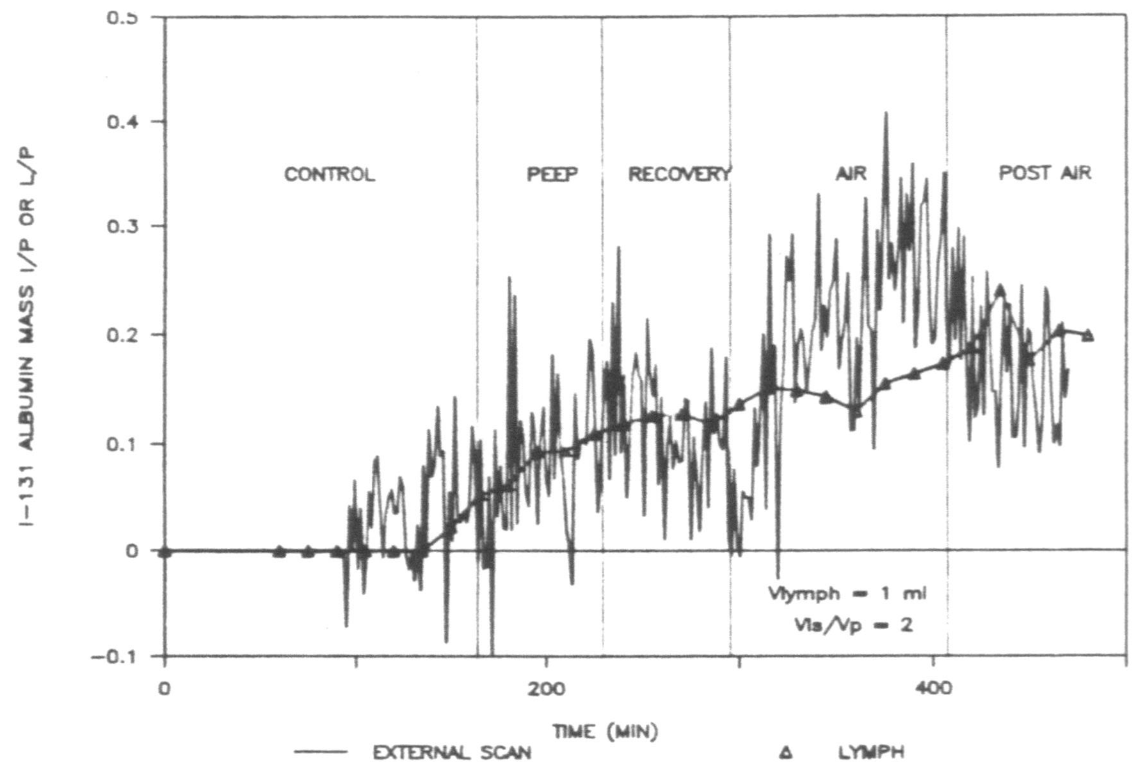

Figure 6. Comparison of I-131 Albumin Mass I/P from external scan data of Figure 5 and from direct measurement in lung lymph. External scan data is corrected for blood volume changes and lymph data is corrected for transit time through the lymphatics.

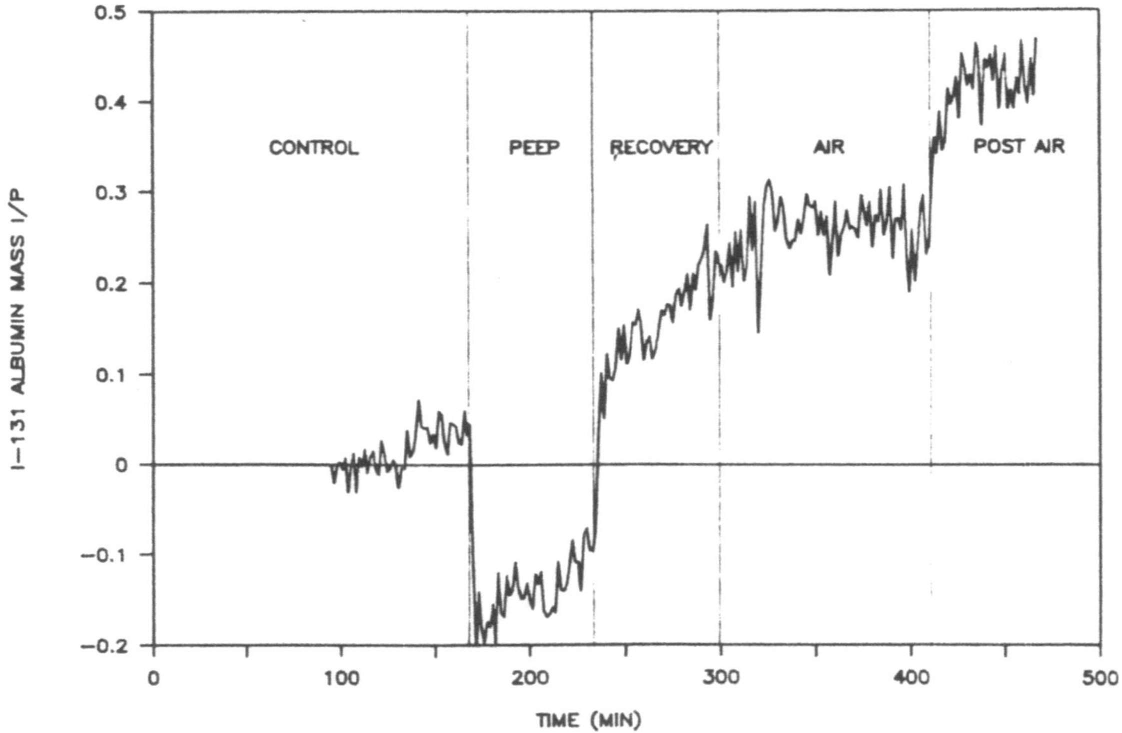

Figure 5. I-131 Albumin Mass I/P ratio following a 15 cm H_2O increase in alveolar pressure and after air a 2 ml/min infusion of air into the pulmonary artery. I/P values are uncorrected for changes in blood volume.

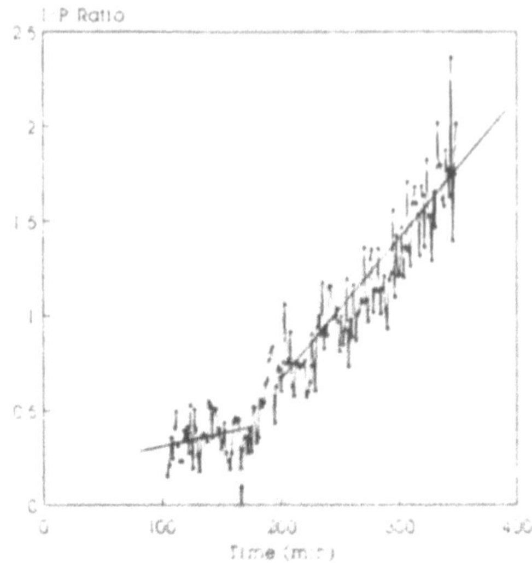

Figure 7. I-131 Albumin Mass I/P in response to an infusion of 20 μg/kg perilla ketone over a 15 minute period. The arrow indicates the start of the infusion period.

The slope of the I-131 concentration vs. time curve is not significantly altered from baseline after microvascular pressure is elevated by 15 cm H_2O (Figure 4). Figure 5 shows the I/P ratio after alveolar pressure is elevated by 15 cm H_2O (PEEP). A jump in I/P is seen in Figure 5, in this figure because it was not corrected for changes in blood volume. The protocol used in these latter studies (Figure 5) was to first establish a 2 hr control period, follow that by a 1 hr recovery period, then infuse air directly into the pulmonary artery at a rate of 2 ml/min for two hours, and finally follow the response for another hour. The data from Figure 5 is corrected for blood volume and compared with the direct measurement of I/P in Figure 6. Agreement is good during baseline, PEEP, and recovery periods. However, the two methods do not agree during air infusion. This may be the result of a redistribution of blood flow or change in interstitial volume during air infusion.

Perilla ketone is known to increase the permeability of sheep lungs (12). The effect of infusing 20 μg/kg of perilla ketone into the in situ sheep lung lymph preparation is shown in Figure 7. Perilla ketone significantly increases the I/P slope, and this is found to be dose-dependent. It should be noted that the maximum value of I/P reached in Figure 7 is above 2, which is much greater than the maximum value found for any of the conditions where permeability was unaltered (e.g., Figures 2 - 6). A direct comparison between the external detection method and lymph measurements is complicated in this case because fluid rapidly accumulates in the interstitium during perilla ketone infusion. To correct for this, an interstitial tracer would be required.

Summary and Conclusions

In summary, we have tested the external scan method for measuring transvascular protein flux using a pump-perfused sheep lung lymph preparation. Results from the external scan method agree with direct lymph measurements under baseline conditions, after microvascular pressure has been elevated, and after alveolar pressure has been elevated. A large increase in the slope of the I/P vs. time relation is found when perilla ketone, a substance which increases lung microvascular permeability in sheep, is infused.

We conclude that the external scan method provides a minimally invasive assessment of transvascular protein transport, and that this technique is sensitive to changes in microvascular permeability.

References

1. Gorin, A.B., W.J. Weidner, R.H. Demling, and N.C. Staub. Non-invasive measurement of pulmonary transvascular protein flux in sheep. J. Appl. Physiol. 45:225-233, 1978.
2. Gorin, A.B. Noninvasive measurement of regional interstitial volume in sheep. J. Appl. Physiol. 64:1561-1566, 1988.
3. Prichard, J. S. and G. de J. Lee. Measurement of water distribution and transcapillary solute flux in dog lung by external radioactivity counting. Clinical Science 57:145-154, 1979.
4. Prichard, J. S., B. Rajagopalan, and G. de J. Lee. Transvascular albumin flux and the interstitial water volume in experimental pulmonary oedema in dogs. Clinical Science 59:105-113, 1980.
5. Dauber, I.M., W.T. Pluss, A. Van Grondelle, R.S. Trow, and J.V. Weil. Specificity and sensitivity of noninvasive measurement of pulmonary protein leak. J. Appl. Physiol 59:564-574, 1985.
6. Gorin, A.B., J. Kohler, and G. DeNardo. Noninvasive measurement of pulmonary transvascular protein flux in normal man. J. Clin Invest. 66:869-877, 1980.
7. Tatum, J.L., T.S. Burke, H.J. Sugerman, A.M. Strash, J.I. Hirsch, and M.J. Fratkin. Computerized scintigraphic technique for the evaluation of adult respiratory distress syndrome:initial clinical trials. Radiology 143:237-241, 1982.
8. Roselli, R.J., R.E. Parker, W.R. Riddle and N.A. Pou. Fluid balance in sheep lungs before and after extracorporeal perfusion. J. Appl. Physiol., In press.
9. Roselli, R.J. and W.R. Riddle. Analysis of noninvasive macromolecular transport measurements in the lung. J. Appl. Physiol. 67:2343-2350, 1989.
10. Riddle, W.R., R.J. Roselli, and N.A. Pou. Modeling flux of free and protein-bound radioisotopes into the pulmonary interstitium. J. Appl. Physiol., In press.
11. Pou, N.A., R.J. Roselli, R.E. Parker, J.A. Clanton, and T.R. Harris. Measurement of lung fluid volumes and albumin exclusion in sheep. J. Appl. Physiol. 67:1323-1330, 1989.
12. Coggeshall, J.W., P.L. Lefferts, M.J. Butterfield, G.R. Bernand, F.E. Carroll, N.A. Pou, and J.R. Snapper. Perilla ketone: A model of increased pulmonary microvascular permeability and pulmonary edema in sheep. Am. Rev. Respir. Dis. 136:1453-1458, 1987.

USE OF MATHEMATICS IN ASSESSING SOLUTE EXCHANGE ACROSS THE LUNG EPITHELIUM

B.T. Peterson, M.L. Collins, J.C. Connelly, J.W. McLarty, D. Holiday and L.D. Gray, University of Texas Health Center at Tyler, Tyler TX 75710 USA

Our early attempts to understand the role of the lung epithelium in the pathophysiology of lung edema and alveolar flooding were hampered by a lack of techniques for specifically assessing changes in lung epithelial permeability to proteins. Consequently, we have developed two methods for assessing solute exchange across the lung epithelium: 1) the use of nuclear imaging techniques to measure the rate of clearance of aerosolized ^{99m}Tc-labeled tracers from the air spaces of the lungs, and 2) the analysis of the clearance of these tracers via the lung lymphatics. Both of these methods employ mathematical modeling techniques to obtain the maximum amount of information from the data.

1. Analysis of Clearance of Radiolabeled Tracers From the Air Spaces of the lungs.

1.1 Assessment of lung epithelial injury

The lung epithelium is a tight barrier that separates the air spaces of the lungs from the lung tissue and alveolar capillaries. Unlike the relatively permeable vascular endothelium, the epithelium normally restricts the movement of proteins between the interstitial spaces and the air spaces (12). However, in the presence of severe lung injury the integrity of both the endothelial and epithelial barriers is compromised and excess fluid and proteins enter the interstitial and air spaces of the lungs creating interstitial and alveolar edema. Clinically, this condition is known by several names, the most common one being the Adult Respiratory Distress Syndrome (ARDS). Injury to the lung endothelium is unquestionably an important part of the pathophysiology of ARDS but the role of the lung epithelium is not clear.

In an attempt to non-invasively assess changes in lung epithelial permeability to solutes, Rinderknecht and coworkers described a nuclear imaging technique to measure the rate of clearance of aerosolized ^{99m}Tc-diethylenetriamine pentaacetate (DTPA, mw=492) from the air spaces (11). Subjects inhaled an aerosol of DTPA for 2-6 minutes and then the radioactivity in the chest was monitored for the next 30-60 minutes. They found that in healthy subjects, DTPA cleared at a rate of 1.6%/min but that in the presence of interstitial lung disease the clearance rate increased to 4.1 %/min (11). Unfortunately, a simple increase in lung volume also increased the clearance rate of DTPA to 2-4 %/min (2). Consequently, it is difficult to interpret an increased DTPA clearance rate because it could be due to either severe lung injury or simple and benign lung inflation (7).

In an attempt to improve this DTPA clearance technique for assessing lung epithelial injury we analyzed the data with the help of 1-,2-, and 3-compartment models (5,8). We also measured the tracer concentration in the lung lymph (4,9) and investigated the use of albumin (mw=70,000) rather than DTPA as a tracer molecule (8,9).

1.2 Compartment analysis

After depositing the aerosolized ^{99m}Tc-labeled tracer in the lungs of anesthetized, mechanically ventilated sheep, we monitored the activity in the chest with a gamma camera programmed to collect the data at 2-minute intervals for a total of 120 minutes. The data were corrected for radioactive decay ($t_{1/2}$=6 hours) and normalized to the peak value that occurred at the end of the deposition period (100%). The time was adjusted so the peak value corresponded to time=0. A curve-fitting subroutine was used to describe the data with a 1-, 2-, and 3-compartment models:

1-compartment: $\text{activity}(t) = 100\, e^{-kt}$

2-compartment: $\text{activity}(t) = A_1 e^{-k_1 t} + (100-A_1) e^{-k_2 t}$

3-compartment: $\text{activity}(t) =$

$$A_1 e^{-k_1 t} + A_2 e^{-k_2 t} + (100-A_1-A_2) e^{-k_3 t}$$

1.3 Selecting the number of compartments

The use of models with more compartments allowed us to more precisely describe the data with a smaller root-mean-square (RMS) error but it was not always clear whether the use of a more complicated model was physiologically or statistically justified. To aid in selection of the most appropriate model, we performed a "runs test" on the residual errors of the data for each model to look for nonrandomness in the fit (8). A "run" is a string of successive residuals, ordered in time, that are all positive or all negative. If there are relatively few runs and the runs are long, the residuals are not randomly distributed. When the residuals obtained from fitting the data to a specific model were reasonably random ($p>0.05$), we considered the model sufficient and the use of more complicated models was not statistically justified. In practice, none of the data we collected required a 3-compartment model (5,8). In addition, we generally found that when the 1-compartment model was insufficient, its RMS error was > 1%. When the runs test showed that the model was sufficient, the RMS error was generally < 1%.

We also investigated the use of a nonlinear regression model for the selection of the most appropriate model by examining the 95% confidence interval of the size coefficient of the multiple compartments. When the confidence interval of any compartment in a multi-compartment model included zero, we concluded that the use of that model was not statistically justified and accepted the model with one less compartment as the most appropriate way to describe the data. We found that this algorithm yielded the same results as those obtained by the runs test (8).

1.4 Physiological meaning of compartments

Once we had determined the best model for describing the clearance data, we had to interpret the physiological meaning of the size and clearance rates of the compartments. The DTPA clearance data in normally ventilated sheep were well described by a 1-compartment model with a single rate constant which we interpreted as the rate of clearance of the DTPA across the lung epithelium.

When the sheep's lungs were inflated to a larger lung volume by the application of 10 cmH_2O positive end expired pressure, a 2-compartment model was required to describe the data. It showed that 56% of the DTPA cleared rapidly and the remainder cleared at the same rate as that measured in the control sheep (5). We suspected that the fast compartment represented the DTPA that had deposited in the alveolar spaces in which surface area greatly increases with lung inflation. Conversely, the remainder of the DTPA, which cleared at the control rate, represented the fraction of the DTPA that deposited in the airways whose surface area is minimally affected by changes in lung volume. This interpretation is supported by our recent study which showed that increasing the aerosol size to minimize deposition in the alveoli decreased the size of the fast compartment (10).

A different interpretation of the compartment analysis was required when we used ^{99m}Tc-human serum albumin (HSA) rather than DTPA as the tracer because the rate of clearance of HSA is essentially unaffected by changes in lung volume (8). The HSA clearance was well described by the 1-compartment model in the control sheep and the sheep with lung inflation (8). However, the 2-compartment model was required for data obtained from approximately 2/3 of the sheep with lung injury and alveolar edema due to an intravenous infusion of oleic acid. In these sheep, 5-15% of the HSA cleared rapidly while the remainder cleared at control or slightly above control rates. We interpreted this small fast compartment to represent the fraction of the aerosolized HSA that had deposited on severely injured regions of the epithelium. This interpretation is supported by gross and microscopic examination of the injured lung which shows a patchy type of lung injury with areas of healthy lung adjacent to areas with severe lung injury (8).

The interpretation that the development of 2 compartments in the clearance of HSA is due to variations in the distribution of epithelial injury is supported by our data with Pseudomonas elastase. When sheep were ventilated with an aerosol of Ps. elastase the clearance rate of HSA increased 6-fold but it was usually well described by a 1-compartment model (6). This would be expected if the aerosol delivery of the elastase caused a uniform increase in permeability throughout all of the epithelial surface area.

2. Compartment Analysis of Lung Lymphatic Clearance of Radiolabeled Tracers From the Air Spaces

2.1 The model

Most of the DTPA and HSA that clears from the air spaces of the lungs is carried away by the blood. However, our measurements of the concentration of the tracers in the lymph that drains the lung interstitial spaces showed that there was exchange of the tracers between the air spaces and the interstitial spaces of the lungs (4,9). We knew that the DTPA and the HSA in the lymph did not all come from the vasculature because the concentration of the tracers in the lymph collected from sheep with lung injury was up to 6 times the concentration of the tracer in the plasma (9). To describe the dynamics of tracer exchange among the air spaces, the interstitial spaces, and the plasma, we used our data to estimate rate constants in a mathematical model of the lungs. The model is shown in the following figure where A(t) is the concentration of the HSA in the epithelial lining fluid of the air spaces, IS(t) is the concentration of the HSA in the interstitial spaces and lung lymph, and P(t) is the HSA concentration in the plasma. K_{alv} is the clearance rate of the HSA from the air spaces into the plasma. However, K_{alv} is not included in the mathematical analysis of the model because P(t) could be measured exactly. K_{epi} and K_{endo} are the rate constants describing the exchange between the interstitial spaces (lymph) and the air spaces and plasma respectively. Q is the measured lymph flow and V (ml) is the interstitial volume drained by the lymphatics. V was estimated from previous indicator dilution studies in sheep and from the body weight of the sheep (3).

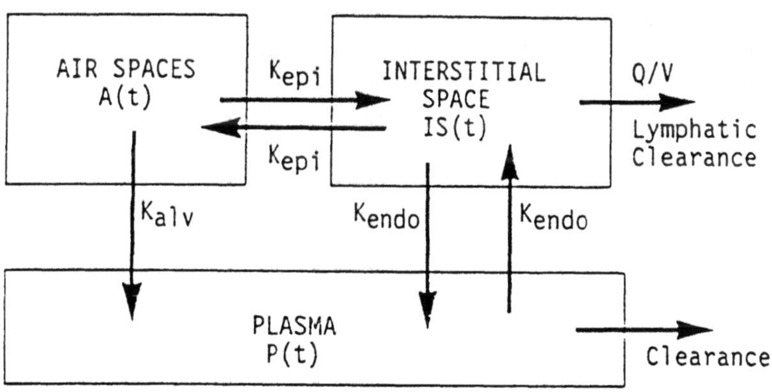

Mathematical model of clearance of tracers form the airspaces of the lung. (Reproduced with permission (9).)

2.2 Physiological interpretation of the model

A kinetic analysis of the HSA movement in this model yields the following expression for the steady-state concentration of the the HSA in the lymph (L):

$$L = \frac{K_{epi} + K_{endo} \times P(t)}{(K_{epi} + K_{endo} + Q/V)}$$

Clearly, the concentration of HSA we measure in the lymph reflects a complicated balance among the exchange rates across the epithelium and endothelium, the plasma concentration, and the lung lymph flow. This is not new information, but by rearranging the equation we can obtain an expression for the rate constant for the exchange of HSA between the air spaces and the lung tissue drained by the lymphatics:

$$K_{epi} = \frac{L(K_{endo} + Q/V) - K_{endo} \times P}{(1-L)}$$

Using previously published measurements of K_{endo} in control sheep and sheep with lung injury (13), we used this expression to estimate K_{epi} from measurements of the lymph flow (Q), and HSA concentrations in the lymph (L) and plasma (P) of anesthetized sheep with lung injury(9).

2.3 Applications to Lung Injury

We found that in anesthetized sheep, acute lung injury increases K_{epi} approximately 80-fold from a control value of 0.06×10^{-5}/min to 4.70×10^{-5}/min (9). This suggests that lung injury causes more of the epithelium to participate in the clearance of the HSA from the air space into the lung tissue drained by the lymphatics. This shows that the movement of protein between the lung interstitium and the air spaces can increase by 80-fold in the presence of lung injury. If, as some have suggested, alveolar flooding is the result of fluid entering the air spaces from the interstitial spaces, then an 80-fold increase in the clearance rate by this route could be a major factor in the pathogenesis of alveolar flooding (1).

3. References

1) Gee, M.H., and N.C. Staub. Role of bulk fluid flow in protein permeability of the dog lung alveolar membrane. J. Appl. Physiol. 42: 144-149, 1977.

2) Marks, J.D., J.M. Luce, N.M. Lazar, J. Ngao-Sun Wu, A. Lipavsky, and J.F. Murray. Effect of increases in lung volume on clearance of aerosolized solute from human lungs. J. Appl. Physiol. 59: 1242-1248, 1985.

3) Peterson, B.T., T.R. Harris, and K.L. Brigham. Comparison of sodium and urea as indicators of pulmonary vascular permeability. Exp. Lung Res. 4: 79-92, 1983.

4) Peterson, B.T., and L.D. Gray. Pulmonary lymphatic clearance of ^{99m}Tc-DTPA from air spaces during lung inflation and lung injury. J. Appl. Physiol. 63: 1136-1141, 1987.

5) Peterson, B.T., H.L. James, and J.W. McLarty. Effects of lung volume on clearance of solutes from the air spaces of lungs. J. Appl. Physiol. 64: 1068-1075, 1988.

6) Peterson, B.T., M.L. Collins, and A.O. Azghani. Aerosolized Pseudomonas elastase increases lung epithelial permeability to albumin in anesthetized sheep. Am. Rev. Respir. Dis. 139: A294, 1989 (Abstract).

7) Peterson, B.T. Pulmonary clearance of aerosolized ^{99m}Tc-DTPA and the lung epithelium. J. Aerosol Medicine 2: 315-328, 1989.

8) Peterson, B.T., K.D. Dickerson, H.L. James, E.J. Miller, J.W. McLarty, and D.B. Holiday. Comparison of three tracers for detecting lung epithelial injury in anesthetized sheep. J. Appl. Physiol. 66: 2374-2383, 1990.

9) Peterson, B.T. and K.D. Dickerson. Concentration of aerosolized ^{99m}Tc-albumin in the pulmonary lymph of anesthetized sheep. J. Appl. Physiol. 68: 1233-1240, 1990.

10) Peterson, B.T., M.L. Collins, L.D. Gray, and D. Suez. Effect of aerosol size on the clearance of DTPA from lungs of anesthetized sheep. J. Aerosol Medicine. In Press.

11) Rinderknecht, J., L. Shapiro, M. Krauthammer, G. Taplin, K. Wasserman, J.M. Uszler, and R.M. Effros. Accelerated clearance of small solutes from the lungs in interstitial lung disease. Am. Rev. Respir. Dis. 121: 105-117, 1980.

12) Taylor, A.E., and K.A. Gaar. Estimation of equivalent pore radii of pulmonary capillary and alveolar membranes. Am. J. Physiol. 218: 1133-1140, 1970.

13) Vaughn, T.R., Jr., A.J. Erdman III, K.L. Brigham, W.C. Woolverton, W.J Weidner, and N.C. Staub. Equilibration of intravascular albumin with lung lymph in unanesthetized sheep. Lymphology 12: 217-223, 1979.

Acknowledgements

The authors gratefully acknowledge the support from the National Heart, Lung and Blood Institute Grant HL-34538 and the Texas Affiliate of the American Heart Association Grant 86G-594.

ON LINE COLORIMETRIC DETERMINATIONS OF TRANSVASCULAR FLUID AND PROTEIN TRANSPORT IN ISOLATED LOBES. L. Oppenheimer, E. Furuya, K.P. Landolfo and D. Huebert. Department of Surgery, University of Manitoba. Winnipeg, Canada.

A commonly used method to estimate transvascular fluid exchange in organs is based on measurements of weight changes with time. Following changes in pulmonary capillary hydrostatic pressure (Pc), weight changes in characteristic fashion. A fast change in weight, assumed to represent vascular volume adaptations, is followed by slow exponential changes, which are thought to represent transvascular fluid exchange associated with the change in Pc. The analysis of the slow exponential weight change constitutes one of the most common approaches to determinations of vascular membrane and transvascular force readjustment in isolated lung preparations such as that depicted in Figure 1.

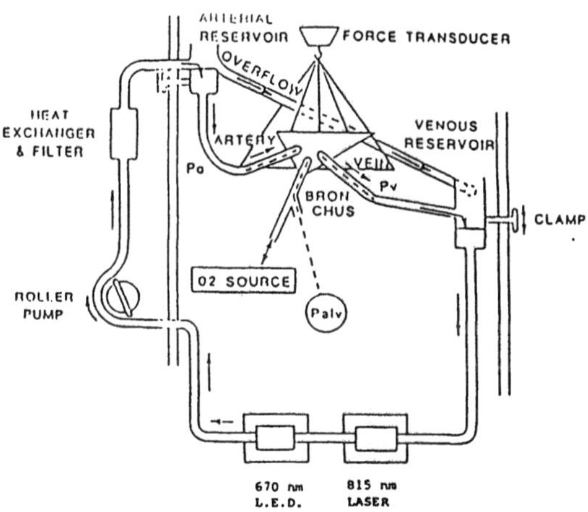

Figure 1

Excised pulmonary lobes are suspended from a force transducer for continuous weight estimates. They are perfused with blood. Pc is changed with adjustmens in the reservoirs. In this preparation, we have interposed a colorimetric device capable of obtaining very accurate determinations of on line hematocrit (1). This device continuously measures the proportion of 815nm light emitted by a 100mW CW laser diode, which is transmitted across the column of perfusate. We have demonstrated that there is a very good correlation between the logarithm of the signal proportional to light transmission and hematocrit (1). We have also shown that changes in hematocrit (Hct) correspond very well with the volume of fluid exchanged with the lobe (V_E) according to the following equation:

$$V_E = C_V \cdot (1 - Hct_I / Hct_{(t)}) \qquad (1)$$

where C_V represents the initial circuit volume, and Hct_I and $Hct_{(t)}$ represent initial and on line changes in Hct at any given time (t). A representative experiment is summarized on Figure 2. Weight changes (dashed line) demonstrate the characteristic fast and slow exponential behavior.

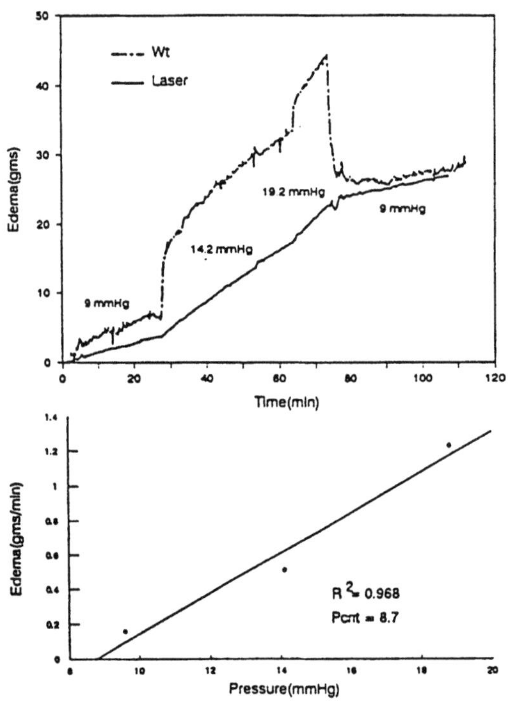

Figure 2

However, estimates of V_E using the colorimetric device (solid line) demonstrate transvascular exchange at constant rates proportional to Pc. Based on these observations we concluded that exponential weight changes reflected not only transvascular fluid exchange, but also previously unrecognized slow vascular volume changes (1).

In addition to the hydrostatic pressure difference across the vascular membrane, transvascular fluid exchange is also influenced by the transmural oncotic gradient, which is the result of the protein concentration differences across the vessel wall. It would be therefore advantageous to develop a method to estimate determine the transvascular protein transport. We have attempted to do so with the addition of a second colorimetric device, and labelling albumin in the perfusate with Evans Blue (EB) at a final concentration of 4mg/dl. EB does not affect 815 nm light transmission. The second device is also illustrated in Figure 1 (670 nm). This second device generates a signal proportional to light absorption of this wavelength by blue dyes. This wavelength is emitted by 5 powerful light emitting diodes (LED) placed in a row along the tubing and connected in series. The proportion of absorbed light is determined by a large surface detector, which adds the fraction of light transmitted from each individual LEDs. Because the 670 nm device is not only affected by changes in Evans Blue concentration, but also by changes in Hct, it becomes necessary to separate one effect from the other. To determine both transvascular fluid and Evans Blue labelled albumin transport, the following steps are taken:

a).- Determination of 670/815 ratio: This step is directed at separating 670 nm signal changes associated with changes in Hct, from the effects of Evans Blue (EB). After EB has been added to the circuit and has had time to mix well, we obtain a 20 ml sample of blood, which is centrifuged in order to concentrate the red cells. After discarding most EB containing supernatant, the cells are resuspended and the sample reinfused into the circuit. Because the concentration of EB in the suspending supernatant has not changed, signal changes in both devices are exclusively the result of changes in Hct.

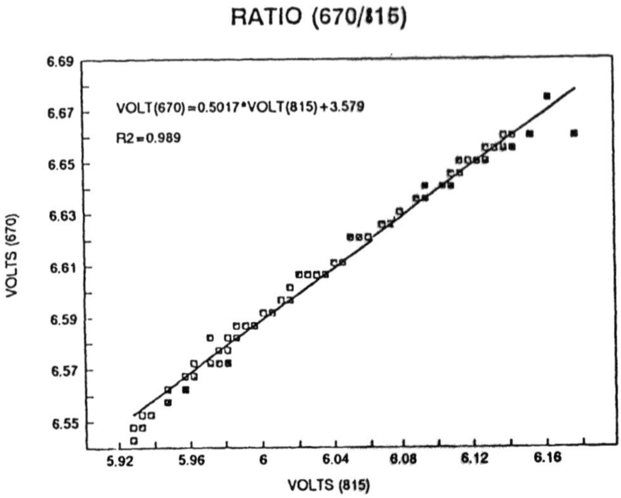

Figure 3

In Figure 3, we have plotted changes registered by the 670 nm detector against 815 nm signal changes. The correlation is excellent. Using the slope of this relationship (R), it is possible to predict changes in 670 nm due to changes in Hct, from simultaneously measured changes in the 815 nm signal.

b).- Calibration of 670 and 815 nm signals. After R has been obtained, we introduced 3 EB free plasma boluses (5 ml each) into the perfusate.

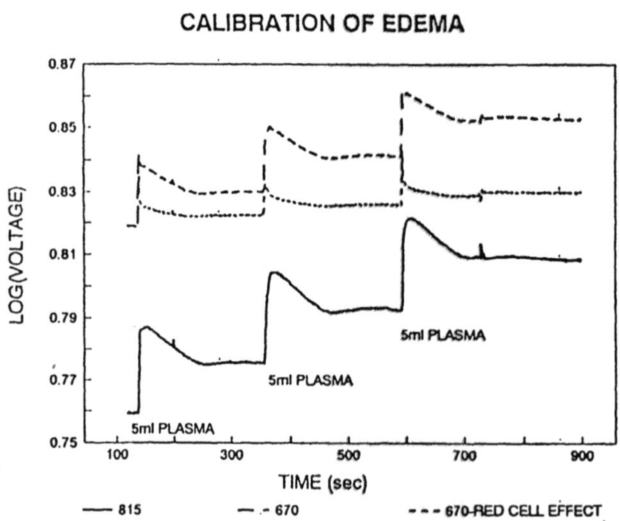

Figure 4

In Figure 4 we have plotted the logarithm of both signals against time. The solid and dashed lines represent the 815 and 670nm signals respectively. Note how after the initial acute deflection, a new steady state is reached after several passes. We used the slope (815cal) of the least squares linear best fit relating volume of plasma added to the change in log(815nm), to convert 815 signal changes to volume exchanged (V_{E815}).

The dotted line represents changes in 670nm light transmission when the red cell (Hct) changes have been subtracted as follows:

$$\Delta\log(670_{EB}) = \Delta\log(670 - R\cdot\Delta 815) \qquad (2)$$

this signal is proportional to changes in EB concentration induced by the EB free plasma boluses. The slope (EBcal) of the least squares linear best fit relating volume of plasma added to change in $\log(670_{EB})$ was used to estimate transvascular volume exchanged (V_{EEB}) from changes in EB concentration.

c).- Determination of transvascular fluid and protein exchange: After calibrations have been obtained Pc is changed in steps to induce transvascular fluid exchange. This is accomplished by changing the height of the reservoirs (Figure 1).

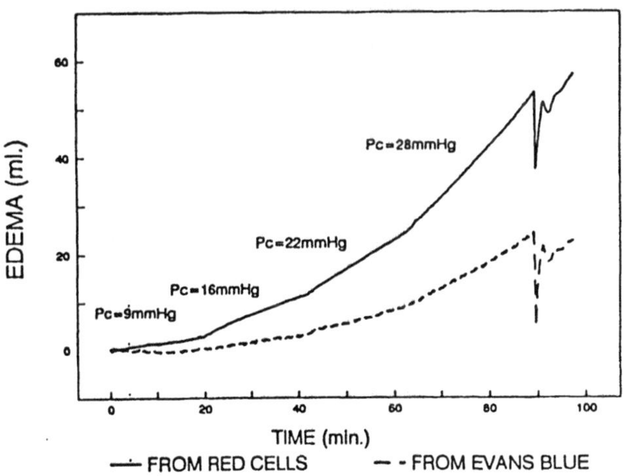

Figure 5

In Figure 5 we are reproducing the results from a typical experiment. Signal changes from both devices were digitized at 1Hz and calculations of V_{E815} (solid line) and V_{EEB} (dashed line) obtained on line using the calibrations previously described. Note how the rate of transvascular fluid exchange (slope of solid line) is always constant and proportional to Pc. The slope of the least squares linear best fit relating the rate of transvascular exchange to Pc represents the vascular membrane conductance or filtration coefficient (Kfc) to fluid exchange. The Pc at which no exchange occurs reflects complete balance of hydrostatic and oncotic forces across the membrane.

Note the discrepancy between V_{E815} and V_{EEB}. V_{EEB} always underestimates V_{E815}. The discrepancy reflects the fact that protein transport occurs across the membrane. Hence EB is not concentrated in the same proportion as red cells. Protein transport as a fraction of its plasma concentration (clearance, Cl_P) can be calculated as follows:

$$Cl_P=(V_{E815}-V_{EEB})/V_{E815} \qquad (3)$$

Determination of Cl_P at various rates of transvascular fluid exchange can be used to elucidate diffusive and convective protein transport (2,3). When Cl_P is constant at increasing rates of transvascular fluid exchange, it represents a measure of the sieving properties or reflection coefficient (σ) of the membrane to albumin. When Oleic Acid (O.A.), a substance capable of inducing a capillary permeability injury was added to the perfusate (0.15ml), Kfc and σ changed from 0.0287±0.0064 $ml \cdot min^{-1} \cdot mmHg^{-1}$ and 0.69±0.31 at baseline to 0.0771±0.0375 $ml \cdot min^{-1} \cdot mmHg^{-1}$ and 0.29±1.5 respectively. The injury increased membrane conductance to transvascular fluid flux and reduced the sieving ability of the membrane, and thus its ability to maintain an effective oncotic gradient. The colorimetric technique using to independent devices has proved useful in characterizing membrane properties, as well as in tracking changes following membrane injury.

REFERENCES:

1.- Hancock, B.J., K.P. Landolfo, M. Hoppensack, and L. Oppenheimer. Slow phase of transvascular fluid exchange reviewed. J. Appl. Physiol. (in press), 1990.

2.- Taylor, A.E., and J.C. Parker. Pulmonary interstitial spaces and lymphatics, In: Handbook of Physiology, Sect. 3, Vol. 1, Chap. 4, Respiration, Washington D.C., American Physiologic Society, 1984, pp. 167-230.

3.- Michel, C.C. Fluid movement through capillary walls, In Handbook of Physiology, Sect. 2, Vol. 4, Chap. 9, The Cardiovascular System, Washington D.C., American Physiologic Society, 1984, pp. 375-409.

Fractal analysis of lung fluid flow

James E. McNamee
Department of Physiology
University of South Carolina School of Medicine
Columbia, SC 29208.

Introduction

The human lung brings a continuous flow of blood into close proximity with a cyclic flow of air so that respiratory gases can readily diffuse between these two fluids. Nature has created an intricate arrangement of spaces to accomplish this task. A 3-dimensional cylinder of venous blood leaving the right ventricle is transformed into a nearly 2-dimensional film of blood by the time it arrives at the alveolus. At the same time, a bolus of inhaled air is divided into smaller streams and pockets until its surface area approaches 100 m^2. Both processes distribute a fluid through a repeatedly bifurcating network. The configuration of these networks as well as their relative sizes have been difficult to summarize using the conventional language of Euclidean geometry [7].

Transforming a 3-dimensional object into something having more nearly a 2-dimensional shape is a feat readily accomplished using fractal geometry. Fractals are formed by repeatedly operating on an object and its constituent parts with ever finer detail. Fractal objects share with lung tissue the property that they appear to lie between two Euclidean dimensions [4]. A fractional dimension, D, descirbes how nearly a fractal object approximates a conventional Euclidean shape.

Dichotomously branching networks form fractal trees that bear striking similarity to lung airway and vascular anatomy [4]. Not surprisingly, mammalian pulmonary bronchial and vascular trees exhibit qualities associated with fractals including self-similarity and scaling [7, 3]. It is reasonable to expect that fractal mathematics and fractal objects can provide more accurate and concise descriptions of pulmonary structure and function.

The fractal tree offers advantages for mass transfer. One benefit of transforming respiratory fluids from 3- to 2-dimensional objects is that diffusive exchange of respiratory gases occurs with great speed. A disadvantage of a tree-like network is that distribution of flow can be nonuniform. Flow through a bifurcating network generally results in an uneven distribution of fluid to peripheral tissue regions [1]. Flow heterogeneity reduces the efficiency of air-blood exchange and can hinder diffusive exchange.

If flow were delivered to each microscopic element of the lung in proportion to tissue mass, tissue perfusion would be constant and uniform throughout the organ. The variance of tissue perfusion obtained by dividing the lung into smaller and smaller pieces and measuring the flow per unit mass of tissue would be zero and it would not increase regardless of the number or sizes of tissue samples analyzed. In contrast, if blood flow were randomly delivered to lung tissue regions, tissue perfusion would be heterogeneous. The variance of tissue perfusion would grow in proportion to the number of tissue samples into which an organ were divided. In living tissue, perfusion is neither homogeneous nor completely random. Its behavior is intermediate, slightly favoring uniformity over disorder. Fractal tree models can provide descriptions of perfusion heterogeneity and can suggest mechanisms which might be involved.

Proportionate effect model

A bifurcating network diverts flow in a parent vessel within one generation into two daughter vessels lying in the next generation. Daughter vessels become parent vessels for future generations of the network. Conservation of mass requires that one daughter vessel receive a fraction γ of parent vessel flow and its sister vessel receive the remaining fraction, $1-\gamma$.

Van Beek, et al. [6] investigated the distribution of flow in bifurcating networks. Their goal was to explain the spatial heterogeneity of myocardial tissue perfusion. They studied several algorithms which allocate flow unevenly to daughter vessels: constantly asymmetric (γ is the same at all bifurcations), progressively asymmetric (γ is reduced by a constant fraction f from one generation to the next) and randomly asymmetric (γ is a Gaussian random variable with constant mean and variance). These strategies are conceptually simple, they assure conservation of mass throughout the network and they produce easily calculated flow patterns. Resulting models produced realistic patterns of blood flow heterogeneity through the first 3-8 generations of network vessels.

The same investigators were able to derive an equation relating the coefficient of variation (relative dispersion, *RD*) of flow to the branching order of the network for the two deterministic models. For the simple case of the constantly asymmetric network,

$$RD(n) = \sqrt{2^n[\gamma^2 + (1-\gamma)^2]^n - 1} \tag{1}$$

where n is the generation within the network. It can be seen that RD increases with n provided that $\gamma \neq 0.5$. Assuming equal masses of tissue are supplied by each branch the rate at which RD grows from one generation to the next determines the fractal dimension of tissue perfusion heterogeneity:

$$D = 1 + \frac{\partial \, Log\, RD}{\partial \, Log\, N} \tag{2}$$

where $N = 2^n$ is the number of vessels in generation n. Since the number of daughter branches doubles between generations, D between successive generations can be rewritten as:

$$D = 1 + \frac{Log \frac{RD(n+1)}{RD(n)}}{Log\, 2} \tag{3}$$

It is enlightening to consider the asymptotic behavior of D for these simple deterministic networks. When γ is near 0.5 and n is small, equation 3 becomes:

$$D = 1 + \frac{Log \frac{n+1}{n}}{2\, Log\, 2} \tag{4}$$

This result means that the fractal dimension of flow heterogeneity in the constantly asymmetric model is initially 1.5 and is independent of the degree of flow asymmetry. This is not caused by network topology but is a consequence of the manner in which flow is diverted from generation to generation. A similar analysis for the progressively asymmetric model indicates its fractal dimension is initially greater than 1.5, its D only slightly depends on flow asymmetry (γ) but it is strongly influenced by the rate, f, at which γ decreases between each generation.

As n becomes very large equation 3 may be approximated by the expression:

$$D = 1 + \frac{Log(1 + 4(\frac{1}{2} - \gamma)^2)}{2\, Log 2} \tag{5}$$

D approaches a limiting fractal dimension which is depends weakly on flow asymmetry. For the constantly asymmetric model, the limiting value of D is always less than 1.03 whenever $0.4 < \gamma < 0.6$. Surprisingly, $D = 1.5$ is the limiting fractal dimension for the progressively asymmetric model.

Proportionate effect models attribute flow heterogeneity primarily to actions occurring at bifurcations. Flow in a parent vessel is made to divide in certain proportions at each bifurcation and fluid proceeds without regard to downstream events. These models produces a Poisson distribution of flows [6]. The shape of the Poisson distribution approaches the shape of the continuous lognormal probability density function as the number of generations grows. West and Shlesinger [8] have shown that a lognormal distribution exhibits fractal-like properties over a limited range but it fails to sustain scaling fractal relationships, in general. The ever changing D in these models may arise from this limitation.

For practical reasons most experimental observations are obtained at small values of n even when the order of the network is large. D is curvilinear in this region and it differs significantly from its limiting value. While a fractal dimension can be computed for constantly and progressively asymmetric models, the dependence of D on n makes interpretation of this measure of perfusion heterogeneity problematic. It also calls in to question the implication of the assumptions upon which these models are based.

Constant difference model

It does not appear that fluid distribution networks in the lung or in most other organs satisfy the assumptions employed by proportionate effect models. The role a vascular network is to provide a supply of flow to peripheral tissues. Except in unusual circumstances, parent vessels do not usually govern the flow to daughter vessels. Instead, demands or constraints imposed by daughter vessels dictate the flow supplied by a parent. This is the approach taken in the vascular beds where the control of flow normally resides with arterioles. An alternative description to the proportionate effect model postulates that control of flow is principally under the influence of peripheral, not central, events [5]. Somewhat different results are achieved when the distribution of flow is controlled by distal branches within a network.

Assume that local requirements or constraints for tissue perfusion are nearly equal in two adjacent tissue elements supplied by two daughter vessels arising from a common parent. The flow in each of these sisters would not be exaxtly equal but would differ by a small amount. Tissue elements supplied by other pairs of sister vessels in the same generation might have somewhat different needs and would receive correspondingly different flows. Flows within other pairs of sister vessels would also be likely to differ from each other by a small amount. As the scale of tissue regions and network vessels changes, both absolute flows and the differences between flows in sister vessels would be expected to change.

The hypotheses that the difference between flows in sister vessels in each generation is nearly constant and that a proportionate change in this flow mismatch from one generation to the next are represented in the modified Weierstrass function. Let the function *W* be a discretized sine function:

$$W(x) = \begin{cases} 1, & \sin(x) \geq 0 \\ -1, & \sin(x) < 0 \end{cases} \tag{6}$$

Then, starting with the most distal vessels in a network of *n* generations, the flow in each branch is given by:

$$Q_n(k) = \frac{Q_0}{2}\sum_{u=0}^{n} a^u\left(1 - W(2\pi b^u \frac{k}{2^n})\right) \tag{7}$$

and

$$Q_{n-1}(k) = Q_n(2k) + Q_n(2k-1) \tag{8}$$

where *k* specifies one of 2^n vessels. $Q_n(k)$ obeys a power-law relation from generation to generation. Flows in sister vessels differ by a constant amount and decrease from one generation to the next. Flow in each parent vessel is the sum of the flows in its two daughter branches.

Figure 1 compares the spatial distribution of flow in a 6th-order constant difference model with $b=2$ to that of a corresponding proportionate effect model. It can be seen that the allocation of flow in each model is similar with small differences apparent at early, middle and later portions of the generation. The models diverge progressively as the order of the network increases.

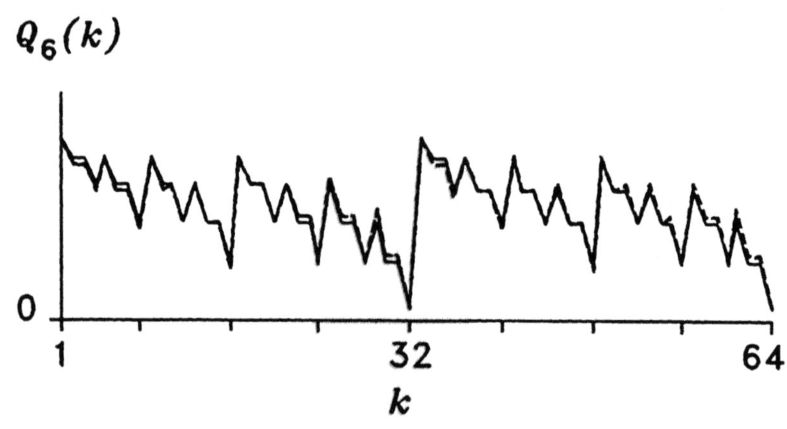

Fig. 1 Comparison between flows through 6th generation vessels in the constant difference (solid line) and proportionate effect (dashed line) models.

When the heterogeneity of flow in this model is analyzed it is found that *RD* increases with generation number but, unlike the case of a proportional distribution model, *D* remains constant over all generations:

$$D = 1 + \frac{\log a}{\log 2} \tag{9}$$

The differences between flows in sister vessels increases with increasing *D* and decreases with increasing generation number. This behavior is consistent with observed flow heterogeneity.

Discussion

Fluid distribution throughout an organ is inherently a nonuniform process. The degree of nonuniformity is found to increase as measurements are made on a finer scale. Empirical determinations of blood flow heterogeneity in the heart [1] and more recently in the lung [2] suggest that heterogeneity of organ perfusion behaves in a fractal manner. A fractal dimension of 1.2 appears to categorize this aspect of flow heterogeneity for both the pulmonary and myocardial circulations.

The physiological mechanisms responsible for the fractal behavior of flow dispersion remain unknown. It is conceivable that vessel length, radius or angle of bifurcation of the first several generations of a network are primarily responsible for establishing the pattern of flow in distal vessels. Central effects can exert a strong influence on the spatial distribution of tissue perfusion. Emboli and atherosclerotic plaques in proximal blood vessels can cause redistribution of flow to the periphery. However, under normal circumstances central vessels do not usually exert much influence on flow.

In addition to the influence of network topology on flow distribution, most organs and tissues exercise some degree of peripheral control over the magnitude and spatial distribution of flow. Blood flow to tissue is proportional to demands and constraints of the local tissue environment under many circumstances . For example, hypoxic vasoconstriction diverts pulmonary blood flow away from poorly ventilated regions of the lung and towards more effective gas exchanging units. Local mechanisms arising in peripheral portions of the distribution network can minimize or exaggerate flow heterogeneity within a topographical region. Flow heterogeneity arising from peripheral portions of a physiological network can be more than a passive geometric phenomenon. Structural aberrations in proximal portions of a bifurcating network may result in little disruption of tissue perfusion when an active peripheral mechanism is capable of adjusting downstream flow.

A description of network flow heterogeneity must incorporate the possibility of peripheral mechanisms which redistribute flow. A new model has been proposed to explain the fractional dimension of tissue perfusion heterogeneity. It is based on a scaling relationship which assumes that flow mismatch in sister branches of a distribution network remains relatively constant within each generation but that flow mismatch will change in a proportionate manner between generations. A modified Weierstrass function expresses these assumptions. The approaches taken in this model provide a role for control processes arising from peripheral sites within a flow distribution network. The model relies on topographical information about the network, in this case the bifurcating nature of the distribution system, only to reconstruct the supply of flow in parent vessels to peripheral tissue elements.

References

[1] Bassingthwaighte, J. B., R. B. King and S. R. Roger. Fractal nature of regional myocardial blood flow heterogeneity. *Circ. Res.* 65:578-590, 1989.

[2] Glenny, R. and H. T. Robertson. Fractal modeling of pulmonary blood flow heterogeneity. *FASEB J.* 4:A1253, 1990.

[3] Lefevre, J. Teleonomical organization of a fractal model of the pulmonary arterial bed. *J. Theor. Biol.* 102:225-248, 1983.

[4] Mandelbrot, B. B. *The fractal geometry of nature.* San Francisco, W. H. Freeman, 1983.

[5] McNamee, J.E. Fractal dimension of tissue fluid flow dispersion. XIV European Conference on Microcirculation, Zurich, (in press), 1990.

[6] Van Beek, J. H. G. M., S. A. Roger and J. B. Bassingthwaighte. Regional myocardial flow heterogeneity explained with fractal networks. *Am. J. Physiol.* 257 (*Heart Circ. Physiol.* 26): H1670-H1680, 1989.

[7] West, B. J., V. Bhargava and A. Goldgerger. Beyond the principle of similitude: renormalization in the bronchial tree. *J. Appl. Physiol.* 60: 1089-1097, 1986.

[8] West, B. J. and M. Shlesinger. The noise in natural phenomena. *Amer. Sci.* 78: 40-45, 1990.

The Roles of Small Molecules as Probes of Endothelial Barrier Function in the Lung: Novel Measurement Methods and Molecular Probes

Thomas R. Harris, Departments of Biomedical Engineering and Medicine, Vanderbilt University, Nashville, TN 37232.

Background

Dysfunction of the barrier properties of endothelial cells is implicated in a number of vascular diseases. One of the more prominent problems is the edema of the lungs seen with Adult Respiratory Distress Syndrome (1,2). Fluid accumulates in the lungs of such patients even though pulmonary pressures are relatively normal. Experimental and patient studies suggest that the primary defect is an increased permeability of the lung vascular capillaries to fluid and macromolecules (3,4,5,6). A variety of methods have been used to assess such malfunctions in experimental animal models of ARDS and in patients. These range from straightforward measurements of lung water by post mortem desiccation and weighing to experimental measurement of lung lymph flow and analysis of protein content as compared to plasma protein concentration.

Two methods have been used to quantify transport across the lung endothelial barrier in intact animals and patients. One method involves scanning of the chest for the activity of gamma emitting or positron emitting radioisotopes which have been used as tags of proteins that can escape from the vasculature of the lung into the interstitium (7,8). This method is quite useful in that it is relatively noninvasive and sensitive to increases in macromolecular permeability. However, it requires a relatively long period of time for a single measurement. In addition, the theoretical separation of diffusion and convection across the endothelium is difficult and leads to some ambiguity in the differentiation of high pressure and high permeability pulmonary edema. Further, the interpretation may be complicated by changes in intravascular volume in the lung and losses of label from the tracer macromolecules (9).

Alternative techniques rely on the escape of small molecules from the circulation. Small molecules have been used as probes of lung vascular function for many years. Trace amounts, usually labelled with radioactive atoms, have been injected into the right heart lumen and sampled from the systemic arterial blood. When the injection is rapid and the bolus contains several different markers, this method is called the multiple indicator-dilution technique and was originated by Chinard (10). A diagram of this technique is shown in Figure 1.

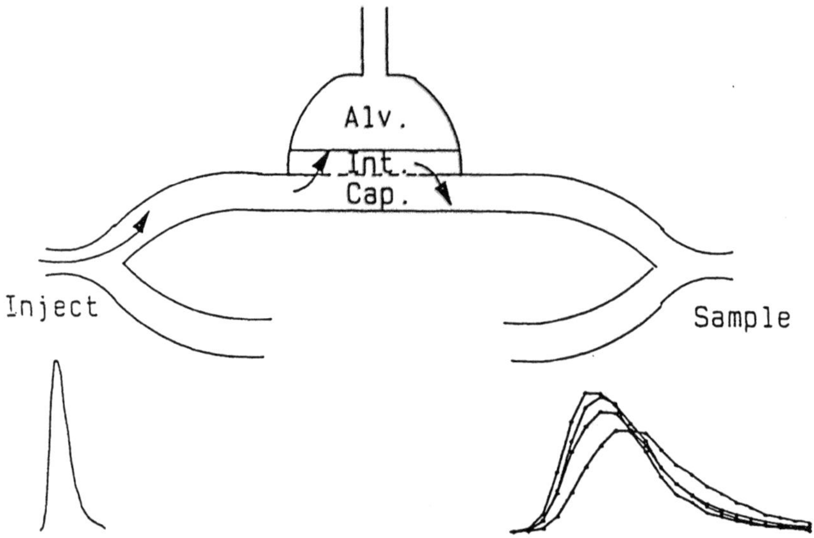

Figure 1. Diagram of the lung multiple indicator dilution technique. The four curves in the right hand side are experimental concentration-time curves of ^{51}Cr-erythrocytes, ^{125}I-albumin, ^{14}C-urea and ^{3}H-water observed after flow through an isolated blood perfused dog lung.

Current interest in the measurement of lung vascular damage preceding respiratory distress syndrome has renewed research into the basic features of the method as a technique for investigation and as a potential tool for clinical assessment (11). Recent research has established that the indicator-dilution method has a number of distinct advantages for evaluating lung vascular function. These are: 1) The separation of a diffusing tracer curve from its appropriate intravascular reference is a direct observation of microvascular transport and therefore must contain information about the state of the microcirculation at the time of indicator passage (11,12). 2) It is a rapid measurement, taking only 30-40 sec, and therefore has the potential to track dynamic changes in the microcirculation (13). 3) It has been shown to quantitatively measure extravascular lung water, microvascular permeability-surface area product, and parameters which characterize saturable uptake by the endothelium (11,13,14,15,16). 4) Parameters derived from indicator curves are altered by lung vascular damage in animal experiments (4,17,18,19). 5) It can be performed in patients under intensive care and provides measures of microvascular function which alter with severity of respiratory distress (6,20,21).

The method has several disadvantages. The blood must be sampled for radioactivity. As in the scanning methods, permeability and surface area are coupled. The method provides a flow weighted average of lung vascular properties. Areas with low flow will make no contribution to the observations. Recent research has tried to address two of these limitations in very specific ways.

Separation of Permeability and Surface Area

The coupling of permeability and surface area in small molecule transport can be derived from a mass balance on unsteady-state exchange of tracer flowing through a capillary as shown in the following equation:

$$\frac{\partial C_D}{\partial t} + u\,\frac{\partial C_D}{\partial x} = -\sum_i N_i \qquad (1)$$

where C_D is the concentration of diffusing tracer in the capillary, N_i are parallel pathways of escape through the capillary barrier, t is time, u is velocity of the phase in which tracer resides within the capillary and x is distance from the capillary entrance.

Various models have been proposed for N_i. One of the simplest is the Crone-Renkin model (22,23) in which

$$N_i = PS_i\, C_D \qquad (2)$$

Where PS_i is the permeability surface area product for the transport of the diffusing tracer through the ith pathway.

One way to correct for variations in surface area would be to find a pair of tracers for which the sensitivities to permeability and surface area changes were different. If one tracer was highly affected by permeability change and the other mainly altered by surface area, then their quotient (PS_1/PS_2) might be a surface area independent measure of permeability. Olson *et al.* (24) have recently undertaken studies which have shown that ^{14}C-urea and ^{14}C-butanediol are such a pair. Butanediol is both hydrophilic and lipophilic (amphipathic) while urea is primarily hydrophilic. Thus, the two tracers will move across the capillary endothelial barrier as indicated in Figure 2. As discussed above, PS for ^{14}C-urea increases when lung vascular permeability is increased. PS for ^{14}C-butanediol also increases, but not as much. The ratio of $PS_{urea}/PS_{butanediol}$ was shown to increase significantly in isolated perfused dog lungs when permeability was increased with alloxan. These changes occurred regardless of lung size. This suggests that a molecule such as butanediol which has a substantial lipid solubility could be mainly influenced by capillary surface area changes. An amphipathic tracer is better for this purpose than a lipophilic molecule because purely lipophilic molecules tend to equilibrate with extravascular space and are therefore not barrier limited.

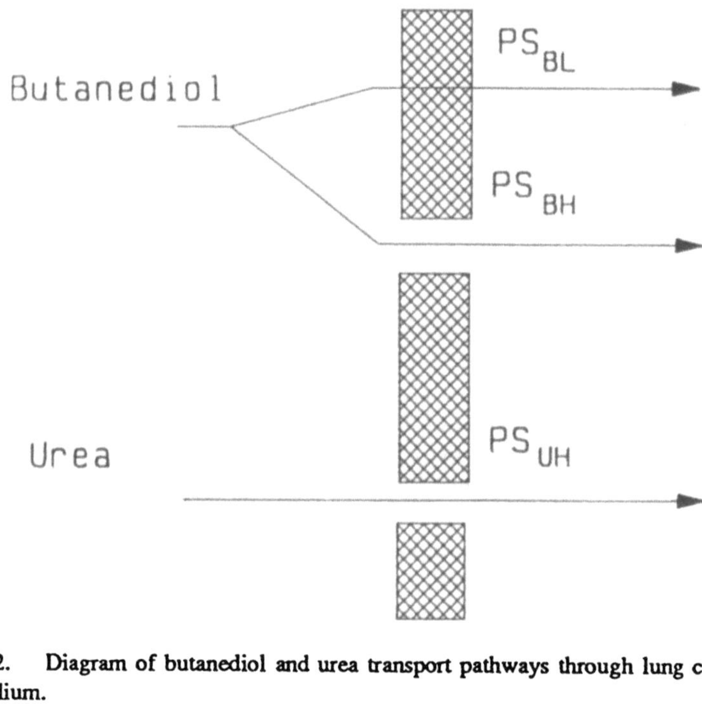

Figure 2. Diagram of butanediol and urea transport pathways through lung capillary endothelium.

Nonradioactive Indicators

Indicator dilution methods based on green dye and other materials capable of absorbing visible light are well established for the measurement of cardiac output. Recently, these methods have been extended to the infrared spectrum by Basset *et al.* (25) and by Neufeld and associates (26). These devices for measuring light transmittance of ICG and D_20 have been connected to microcomputers which calculate mean transit times, pulmonary blood flow and extravascular lung water from lung vascular indicator curves.

These approaches do not measure permeability however. We found that ^{14}C-propanediol had a very similar extraction to that seen for ^{14}C-urea in the isolated perfused dog lung (27). Further, propanediols have a high absorption peak in the near infrared range (9.8 micrometers). Water and blood are relatively nonabsorbent in that range. Therefore we conducted experiments on isolated plasma perfused lungs in which 1,2-propanediol was used as a diffusing tracer and indocyanine green dye was used as an intravascular tracer. The extraction and PS of optically measured propanediol was approximately the same as that found using radioisotopes in the same preparation (28). These results suggest that lung indicator dilution curves can be measured without radioactivity. Further work is needed to extend this technique to whole blood.

Summary

Small molecules have a useful role in the evaluation of lung capillary endothelial function in intact animals and patients. Recent work with amphipathic and optically measurable indicators has reduced some of the complexities inherent in using these methods.

Acknowledgement. This work was supported in part by Public Health Service, National Institutes of Health Grant No. HL19153, SCOR in Pulmonary Vascular Diseases.

References

1. Staub, N.C., Pathophysiology of Pulmonary Edema. In: *Edema*, 30:719-46, Staub and Taylor, eds., New York, Raven Press, 1984.

2. Bernard, G.R., and Brigham, K.L., Pulmonary Edema: Pathophysiologic mechanisms and new approaches to therapy. *Chest*, 89: 594-600, 1986.

3. Esbenshade, A.M., Newman, J.H., Lams, P.H., Jolles, H., and Brigham, K. L. Respiratory failure after endotoxin infusion in sheep: Lung mechanics and lung fluid balance. *J. Appl. Physiol. 53*:967-76, 1979.

4. Bradley, J.D., Roselli, R.J., Parker, R.E., and Harris, T.R. Effects of endotoxemia on the sheep lung microvascular membrane: A two-pore theory. *J. Appl. Physiol. 64*:2675-2683, 1988.

5. Rinaldo, J.E. Borovetz, H.S., Mancini, M.C., Hardesty, R.L., Griffith, B.P. Assessment of lung injury in the adult respiratory distress syndrome using multiple indicator dilution curves. *Am. Rev. Respir. Dis. 133(6)*:1006-10, 1986.

6. Harris, T.R., Bernard, G.R., Brigham, K.L., Higgins, S.B., Rinaldo, J.E., Borovetz, H.S., Sibbald, W.J., Kariman, K., Sprung, C.L. Lung microvascular transport properties measured by multiple indicator dilution methods in ARDS patients: A comparison between patients reversing respiratory failure and those failing to reverse. *Am.Rev.Resp.Dis. 141*:272-280, 1990.

7. Byrne, K., Sugerman, H.J. Experimental and clinical assessment of lung injury by measurement of extravascular lung water and transcapillary protein flux in ARDS: a review of current techniques. *J. Surg. Res. 44*:185-203, 1988.

8. Mintun, Mark A., Dennis, D.R., Welch, M.J., Mathias, C.J., Schuster, D.P. Measures of pulmonary vascular permeability with PET and Gallium-68 transferrin. *J. Nucl. Med. 28*:1704-1716, 1987.

9. Roselli, R.J. and Riddle, W.R. Analysis of non-invasive microvascular macromolecular transport measurements in the lung. *J. Appl. Physiol. 67*:2343-2350, 1989.

10. Chinard, F.P. Estimation of extravascular lung water by indicator-dilution techniques. *Circ. Res. 37*:137-145, 1975.

11. Harris, T.R., Brigham, K.L. The exchange of small molecules as a measure of normal and abnormal lung microvascular function. *Ann. N.Y. Acad. Sci.* *384*:417-434, 1982.

12. Bassingthwaighte, J.B., Goresky, C.A. Modeling in the analysis of solute and water exchange in the microvasculature. *Handbook of Physiology - The Cardiovascular System IV*, American Physiological Society, 1985.

13. Harris, T.R., Bernard, G.R., Roselli, R.J., Maurer, C.R., Pou, N.A. Extravascular lung water by infrared and other measures. *Proc. Ann. Conf. Engr. Med. Biol.* 27:77, 1985.

14. Lewis, F.R., Elings, V.B., Hill, S.L., Christensen, J.M. The measurement of extravascular lung water by thermal-green dye indicator dilution. *Ann. N.Y. Acad. Sci.* *384*:394-410, 1982.

15. Rickaby, D.A., Linehan, J.H., Bronikowski, T.A., Dawson, C.A. Kinetics of serotonin uptake in the dog lung. *J. Appl. Physiol.* *51*:405-414, 1981.

16. Syrota, A., Girauld, M., Pocidalo, J-J., Yudilevich, D.L. Endothelial uptake of amino acids, sugars, lipids, and prostaglandins in rat lung. *Am. J. Physiol.* *243*:C20-C26, 1982.

17. Harris, T.R., Rowlett, R.D., Brigham, K.L. The computation of pulmonary capillary permeability from multiple-indicator data: The effects of increased capillary pressure and alloxan treatment. *Microvascular Research* *12*:177-196, 1976.

18. Harris, T.R., Brigham, K.L., Rowlett, R.D. Pressure, serotonin and histamine effects on lung multiple-indicator curves in sheep. *J. Appl. Physiol.* *44*:245-253, 1978.

19. Zelter, M., Lipavsky, D., Hoeffel, J.M., Murray, J.F. Effect of lung injuries on ^{14}C-urea permeability surface area product in dogs. *J. Appl. Physiol.* *56*:1512-1520, 1984.

20. Brigham, K.L., Snell, J.D., Harris, T.R., Marshall, S., Haynes, J., Bowers, R.E., Perry, J. Indicator dilution lung water and vascular permeability in humans: Effects of pulmonary vascular pressure. *Circ. Res.* *44*:523-530, 1979.

21. Brigham, K.L., Kariman, K., Harris, T.R., Snapper, J.R., Young, S.L. Correlation of oxygenation with vascular permeability-surface area but not with lung water in humans with acute respiratory failure and pulmonary edema. *J. Clin. Invest.* *72*:339-349, 1983.

22. Renkin, E.M. Transport of potassium-42 from the blood to tissue in isolated mammalian skeletal muscles. *Am. J. Physiol. 197*:1209-10, 1959.

23. Crone, C. The permeability of capillaries in various organs determined by use of the "indicator diffusion" method. *Acta Physiol. Scand. 58*:292-305, 1963.

24. Olson, L.E., Pou, A., Harris, T.R. Measurements of amphipatic and hydrophilic indicator-dilution tracers in the lung provide a surface area independent assessment of permeability changes. *FASEB J. 3*:A1140, 1989.

25. Basset, G., Martel, G., Bouchonnet, F., Marsac, J., Sutton, J., Botter, F., Capitini, R. Simultaneous detection of deuterium oxide and indocyanine green in flowing blood. *J. Appl. Physiol. 50*:1367-1371, 1981.

26. Neufeld, G. *et al. Proceedings, 2nd Int. Symposium on Computing in Anesthesia and Intensive Care*, Rotterdam, 1983.

27. Harris, T.R., Roselli, R.J., Maurer, C.R., Parker, R.E., Pou, N.A. Comparison of labelled propanediol and urea as markers on lung vascular injury. *J. Appl. Physiol. 62*:1852-1859, 1987.

28. Galloway, R.L., Jr., Staton, D.J., Harris, T.R. The optical measurement of 1,2-propanediol for the determination of lung capillary permeability surface area. *IEEE Trans. on Biomed. Engr. 36*:591-596,1989.

INTEGRATING MECHANICS AND TRANSPORT IN ASSESSING RESPIRATORY FUNCTION

Michael P. Hlastala

The lungs serve a primary function of exchanging gases between the body and the outside environment. Efficiency of this exchange process is strongly affected by the relative matching of ventilation and perfusion, a relationship which the body goes to great lengths to preserve. However, the efficiency of the lung deteriorates in the presence of lung disease and under conditions of environmental stress. The response of this gas exchange process to stress is largely dependent on the degree of heterogeneity in the lung and the perturbations caused by the pathological or environmental stress.

Lungs exhibit both spatial and temporal heterogeneity. Spatial heterogeneity is caused by different mechanical stresses distributed throughout the lung interacting with the ever-present influence of gravity. Temporal heterogeneity results from the cyclical nature of ventilation and the pulsatile nature of perfusion.

Movement of inspired air is governed by a balance between convection and diffusion. As inspired air progresses down the airways into the lung with increasing airway cross-sectional area, airflow velocity decreases. The decreasing relative importance of convection establishes a stationary diffusion front between the alveolar and inspired gas. It is the position of this diffusion front relative to the branch points of the airways that determines the distribution of inspired gas and thus, the relative efficiency of ventilation.

Inspired air passing through the airways is conditioned before it reaches the delicate alveolar spaces. Heating and humidification is important for processing the air. However, the transfer of energy and water affects the airway mucosa. Bronchoconstriction, bronchial artery dilation and bronchial vascular hyperpermeability are consequences which manifest differently in various species.

Perfusion distribution in the lung is governed primarily by factors influencing the local resistance with little gravitational influence. Local vascular resistance is dependent on local lung volume. Alveolar capillary resistance increases at both low and high lung volume. Extraalveolar vessels decrease resistance as lung volume increases. In addition, corner vessels, which are different from both alveolar and extraalveolar vessels, affect blood flow under certain circumstances. But the physiology of the corner vessels has not yet been investigated systematically. Gravity undoubtedly plays a role, but this now appears to be less important than the local factors. A new approach to dealing with non-gravitational determinants of heterogeneity is fractal analysis, a process that allows characterization of heterogeneity beyond the minimum resolution limits of experimental methodology.

Imperfect matching of ventilation and perfusion is a necessary consequence of the multiple factors which affect ventilation and perfusion nonuniformly. This aspect of respiratory function has been assessed by taking advantage of the differential retention of eliminated gases by pulmonary blood depending on the solubility of that gas. The multiple inert gas elimination technique (MIGET) has provided an excellent tool for physiological and clinical researchers to assess the distribution of V_A/Q in the lung.

This session will explore some of the newer approaches for analyzing the transport properties of the lung. We will consider the implications of heating and humidification of inspired air, non-gravitational factors affecting both ventilation and perfusion distribution and the characterization of V_A/Q heterogeneity with MIGET. These approaches cover only a few of the sophisticated methods available for evaluating lung function. But they give us some insight into the problems faced while integrating mechanics and transport in assessing respiratory function.

AIRWAY HEAT AND WATER EXCHANGE

Julian Solway, Department of Medicine,
University of Chicago, Chicago, IL.

Hyperventilation of dry or cool air causes respiratory heat and water loss, and elicits airflow obstruction in patients with exercise-induced asthma. Though the mechanism by which heat and/or water fluxes are transduced to local airway narrowing remains uncertain, considerable insight has been gained. Theoretical and experimental analyses of intra-airway heat and water transfers have demonstrated that: (1) increasing minute ventilation (MV) promotes further penetration of cool, dry, inspired gas into the lung before it has been fully warmed and humidified to alveolar gas conditions; (2) lowering inspired gas temperature or humidity at fixed MV increases local heat/water losses, but the axial distribution of those losses remains largely unchanged; (3) the net change in airway wall temperature or hydration depends importantly upon the effectiveness of local heat and water replenishing sources that resist the airway cooling and dessication promoted by local heat/water losses. Experimentally documented sequelae of dry gas hyperpnea include bronchoconstriction (found in the central airways of guinea pigs and of human asthmatic subjects, and in the dog lung periphery), broncho-vascular arterial dilatation (confirmed in dogs and sheep, and possibly in normal humans), and broncho-vascular hyperpermeability (demonstrated in guinea pigs and ferrets). One or several facets of the physical events that accompany dry gas hyperpnea lead to the release of bronchoactive and vasoactive mediators (including eicosanoids and tachykinins, from animal studies) that could conceivably provoke airflow obstruction by direct stimulation of airway smooth muscle, or by promoting broncho-vascular engorgement and/or hyperpermeability, with consequent edema formation.

Supported by NHLBI Grants HL02205 and HL41009.

QUANTITATION OF THE REGIONAL DISTRIBUTION OF PULMONARY BLOOD FLOW BY FRACTAL ANALYSIS

H. Thomas Robertson and Robb W. Glenny

1. Introduction

The structure of the pulmonary vascular tree following its initial segmental divisions can be described by a regular recursive algorithm. This vascular structure has fractal properties in that each successive branching is similar to its parent branch over many orders of scale. While the anatomic structure has fractal properties, it remained to be demonstrated that the actual flow distribution of blood within the lung could also be described by a fractal model. We have developed a means of obtaining very high resolution blood flow images of dog lungs, and have demonstrated a previously unsuspected degree of flow heterogeneity. We sought to determine whether the measured heterogeneity of flow (in contrast to the anatomical structure of the pulmonary arteries) could also be described by a fractal model. The method we utilized to demonstrate fractal flow distributions was that developed by Bassingthwaighte and colleagues (1) to describe the heterogeneity of cardiac blood flow.

2. Methods

The heterogeneity of flow in the lungs of normal anesthetized dogs was measured using direct counting of flow labelled lung slices. The flow marker, technetium-99m macroaggregate, was injected intravenously into the supine anesthetized dogs while the lung volume was held at functional residual capacity. The animals were sacrificed by exsanguination under deep anesthesia, a thoracotomy was performed, the pulmonary artery and left atrium were cannulated, and the lungs re-inflated to their FRC volume after closing the thoracotomy incision. The lungs were fixed by intravascular perfusion with glutaraldehyde followed by dehydration with increasing concentrations of ethyl alcohol. The fixed lungs were removed from the thorax, air dried for several hours, and encased in a rapid setting polyurethane foam. The foam block was cut in 11 mm thick transverse slices, and each slice was imaged directly on a gamma camera in the highest resolution mode (pixel size 1.5 x 1.5 mm, measured full width half max, 4.4 mm). The data from the individual slice images were then transferred to a desktop computer and assembled as a three dimensional matrix of 25 mm^3 voxels. (See (2) for additional detail) For six dogs, data sets containing from 2.1 x 10^4 to 5.1 x 10^4 pieces were obtained.

Using the procedure of Bassingthwaighte et al (1), flow heterogeneity was determined as a function of the size of the sample piece considered. The relative dispersion (RD) of blood flow in the lung (RD= standard deviation of flow /mean flow) was calculated for each lung data set. Flow values of pairs of adjacent pieces were then combined, a procedure which doubled the sample piece size and halved the number of pieces for each data set. The RD of the recombined data set was calculated. This procedure of successive doubling of the mean piece size for RD calculation was repeated until a piece volume of roughly 25 cc was reached, and the composite results plotted as ln (RD) versus ln (piece volume).

3. Results:

A linear relationship between ln (RD) and ln (piece volume) was apparent for all dogs over the range of piece sizes considered (Figure 1).

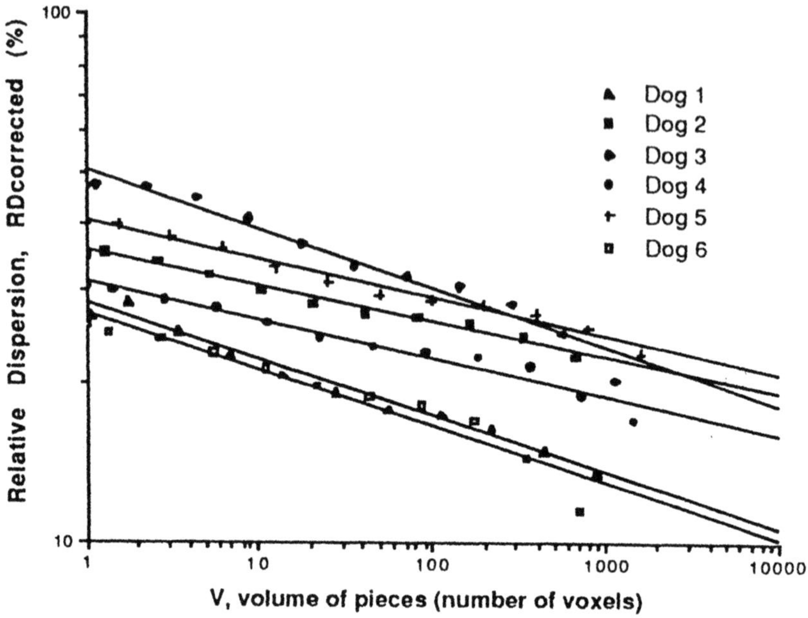

Figure 1. Log-log plot of the relative dispersion of blood flow in the lung calculated as a function of different volumes chosen for flow measurement (data presented in (2)).

This linear relationship on a log-log scale is a necessary condition to describe a process as fractal over the range of the observations made. The slope of the data points on this plot can be related to a spatial fractal dimension, D_S. The equation used to estimate the D_S from the measurements of RD made at different lung piece volumes (v) is given by:

$$RD(v) = RD(v_{ref}) \cdot \left[\frac{v}{v_{ref}}\right]^{1-Ds} \qquad ($$

Where RD(v) is the measured relative dispersion when the or[
is partitioned into regions of a specified volume v, $RD(v_{ref})$ is RD measured for the smallest piece size (v_{ref}), and D_S is

derived fractal dimension. Taking the logarithm of equation 3, we obtain

$$\log RD(v) = (1 - D_s) \cdot \log\left[\frac{v}{v_{ref}}\right] + \log RD(v_{ref}) \qquad (2)$$

If the slope of log(v/vref) is constant over a range of partitions, the system is said to be fractal within that range. The greater the heterogeneity of a system, the greater is the D_S for that system. A D_S of 1.0 reflects a perfectly homogeneous flow distribution, while a D_S of 1.5 would be obtained from a completely random flow distribution (2). The D_S estimate obtained from the six dogs was $1.09 \pm .02$ (mean$\pm$ s.d.). The fractal dimension obtained by this method provides a measure of the scale independent irregularity, roughness, or variation of a system. The advantage of this particular analytical approach is that it provides an estimate of D_S from easily obtained measurements of the RD of regional organ flow during successive subdivisions of the organ pieces down to v_{ref}.

4. Discussion

This study uses a fractal model to demonstrate progressive increases in flow heterogeneity down to a sample size of 25 mm^3. The method described here has a resolution representing the volume of approximately 2500 alveoli. While the distribution of blood flow to the lung will cease to be fractal when the functional perfusion unit of the alveolar capillary is reached, we observed no evidence of a plateau in the log-log plot of RD vs v, which would be expected if the smallest sample size (v_{ref}) were the same size as a functional unit of flow in the lung. The actual heterogeneity of blood flow may therefore be substantially greater than our current methodology can detect. If the alveolus is the functional unit and the slope of the fractal plot is back-projected to a piece size appropriate for an alveolus, then an overall flow RD of 78% would be anticipated. This suggestion of severe flow heterogeneity within the normal lung appears to conflict with observations of the normal overall homogeneity of ventilation-perfusion relationships (3). Two different hypotheses could resolve this apparent contradiction: First, there could simply be very precise local matching of ventilation to perfusion. This precise matching could be based on anatomic and/or local autoregulatory mechanisms. The alternative hypothesis is that there is a far more limited degree of heterogeneity of ventilation (in comparison to perfusion) due to the intrapulmonary gas mixing mechanisms of tidal ventilation, cardiac movement, and collateral ventilation among adjacent lung regions. The assumption which is included as part of this hypothesis is that all capillary blood flow equilibrates with mixed alveolar gas, so that any mixing process which homogenizes alveolar gas over a given region necessarily homogenizes alveolar capillary content and gas exchange within that region. By the gas mixing hypothesis, the heterogeneity of ventilation would be ameliorated, and ventilation-perfusion homogeneity could be maintained even in the face of marked regional perfusion heterogeneity. The two hypotheses cannot be evaluated until both high resolution regional ventilation and high resolution regional perfusion data can be obtained from the same lung.

5. References:

1. Bassingthwaighte JB, RB King, SA Roger. Fractal nature of regional myocardial blood flow heterogeneity. Circ. Res. 65: 578-590, 1989.

2. Glenny RW, Robertson HT. Fractal properties of pulmonary blood flow: Characterization of spatial heterogeneity. J. Appl. Physiol. (In Press) 1990.

3. Hlastala MP, Robertson HT: Inert gas elimination characteristics of the normal and abnormal lung. J. Appl. Physiol. 44(2):258-266, 1978.

ELUCIDATION OF PRINCIPLES OF GAS EXCHANGE BY MEANS OF SOLUBLE TRACER SPECIES

Peter D. Wagner, Department of Medicine
University of California at San Diego, La Jolla, CA. 92093

Because the lung is inaccessible to direct evaluation of events on a fine regional basis in-vivo, its properties are usually inferred from global gas exchange data taken from arterial blood and expired gas. This presentation briefly reviews the many uses of tracer gases of varying physicochemical properties in estimating both global and distributed physiological parameters. In each, the necessary mathematical and physiological underpinnings will be discussed, as will the information content and associated limitations. Approaches to be covered include: (1) Using gases of different molecular weight to infer the importance of rates of diffusion in gas exchange processes that involve both diffusive and convective movement; (2) Using gases of different blood:gas partition coefficients to infer the importance of the convective processes involved in gas exchange, and ventilation/perfusion mismatching, in particular; (3) Using carefully selected gases and procedures to measure variables related to global function (eg, pulmonary blood flow, lung tissue volume, lung gas volume, and lung closing volume); and (4) Using gases with unique chemical properties to determine the importance of chemical rates of reaction between hemoglobin and the gaseous ligands O_2, CO_2, and CO. While all of these techniques have well-known limitations, they have at the same time greatly increased our understanding of how the lung works. When used within the constraints imposed by those limits, they continue to form a group of very useful investigative tools that continue to increase our insights into both normal and abnormal gas exchange processes.

(Supported by HL 17731)

MIX
Papier aus verantwortungsvollen Quellen
Paper from responsible sources
FSC® C105338

If you have any concerns about our products,
you can contact us on
ProductSafety@springernature.com

In case Publisher is established outside the EU,
the EU authorized representative is:
Springer Nature Customer Service Center GmbH
Europaplatz 3, 69115 Heidelberg, Germany

Printed by Libri Plureos GmbH
in Hamburg, Germany